SEXUALITY AND MEDICINE

Volume I

PHILOSOPHY AND MEDICINE

Editors:

H. TRISTRAM ENGELHARDT, JR.

Center for Ethics, Medicine, and Public Issues
Baylor College of Medicine, Houston, Texas, U.S.A.

STUART F. SPICKER

School of Medicine, University of Connecticut Health Center,
Farmington, Connecticut, U.S.A.

VOLUME 22

SEXUALITY AND MEDICINE

Volume I: Conceptual Roots

Edited by

EARL E. SHELP

Institute of Religion, and
Center for Ethics, Medicine, and Public Issues,
Baylor College of Medicine,
Houston, Texas, U.S.A.

D. REIDEL PUBLISHING COMPANY

DORDRECHT / BOSTON / LANCASTER / TOKYO

Library of Congress Cataloging-in-Publication Data

Sexuality and medicine.

(Philosophy and medicine ; v. 22)
Includes bibliographies and index.
Contents: v. 1. Conceptual roots.
1. Sex (Psychology) 2. Medicine and psychology. I. Shelp, Earl E., 1947- . II. Title. III. Series.
BF692.S4347 1987 306.7 86-26201
ISBN 90-277-2290-0 (v. 1)

Published by D. Reidel Publishing Company,
P.O. Box 17, 3300 AA Dordrecht, Holland.

Sold and distributed in the U.S.A and Canada
by Kluwer Academic Publishers,
101 Philip Drive, Norwell, MA 02061, U.S.A.

In all other countries, sold and distributed
by Kluwer Academic Publishers Group,
P.O. Box 322, 3300 AH Dordrecht, Holland.

Printed in The Netherlands.

TABLE OF CONTENTS
Volume I

WILLIAM H. MASTERS AND VIRGINIA E. JOHNSON-MASTERS / Foreword ix

ALAN SOBLE / Preface: Changing Conceptions of Human Sexuality xi

EARL E. SHELP / Introduction xxv

SECTION I: HUMAN SEXUALITY

ROBERT C. KOLODNY / Medical and Psychiatric Perspectives on a 'Healthy Sexuality' 3

FREDERICK SUPPE / Medical and Psychiatric Perspectives on Human Sexual Behavior 17

STEPHEN B. LEVINE / The Origins of Sexual Identity: A Clinician's View 39

LESLIE M. LOTHSTEIN / Theories of Transsexualism 55

VERN L. BULLOUGH / Sex Research and Therapy 73

FRITZ K. BELLER / A Survey of Human Reproduction, Infertility Therapy, Fertility Control and Ethical Consequences 87

SECTION II: SEXUALITY AND SEXUAL CONCEPTS

ALAN SOBLE / Philosophy, Medicine, and Healthy Sexuality 111

JOSEPH MARGOLIS / Concepts of Disease and Sexuality 139

JEROME NEU / Freud and Perversion 153

SANDRA HARDING / The Politics of The Natural: The Case of Sex Differences 185

ROBERT C. SOLOMON / Heterosex 205

ELI COLEMAN / Bisexuality: Challenging Our Understanding of Human Sexuality and Sexual Orientation 225

PETER ROBERT BREGGIN / Sex and Love: Sexual Dysfunction as a Spiritual Disorder 243

NOTES ON CONTRIBUTORS 267

INDEX 268

TABLE OF CONTENTS

Volume II

CRISTIE HEFNER / Foreword ix

NANCY N. DUBLER / Preface xiii

EARL E. SHELP / Introduction xxi

SECTION I: REPRODUCTION, MEDICINE, AND MORALS

MARY ANN GARDELL / Sexual Ethics: Perspectives from the History of Philosophy 3

SARA ANN KETCHUM / Medicine and the Control of Reproduction 17

LISA SOWLE CAHILL / On the Connection of Sex to Reproduction 39

H. TRISTRAM ENGELHARDT, JR. / Having Sex and Making Love: The Search for Morality in Eros 51

SECTION II: SOCIETY, SEXUALITY, AND MEDICINE

JOHN DUFFY / Sex, Society, Medicine: An Historical Comment 69

ROBERT BAKER / The Clinician as Sexual Philosopher 87

FREDERICK SUPPE / The Diagnostic and Statistical Manual of the American Psychiatric Association: Classifying Sexual Disorders 111

JOSHUA GOLDEN / Changing and Medical Practice 137

NELLIE P. GROSE and EARL E. SHELP / Human Sexuality: Counselling and Treatment in a Family Medicine Practice 155

J. ROBERT MEYNERS / Sex Research and Therapy: On the Morality of the Methods, Practice and Procedures 171

SECTION III: RELIGION, MEDICINE AND MORAL CONTROVERSY

PAUL D. SIMMONS / Theological Approaches to Sexuality: An Overview 199

JAMES J. MCCARTNEY / Contemporary Controversies in Sexual Ethics: A Case Study in Post-Vatican II Moral Theology 219

ROBERT H. SPRINGER / Transsexual Surgery: Some Reflections on the Moral Issues Involved 233

RONALD M. GREEN / The Irrelevance of Theology for Sexual Ethics 249

NOTES ON CONTRIBUTORS 271

INDEX 273

FOREWORD

When confronted by the concerns of human sexual function or dysfunction, American medicine finds itself well impaled on the horns of a dilemma. Currently it is acceptable medical practice to treat sexual dysfunctions, disorders, or dissatisfactions that arise from psychogenic etiologies, endocrine imbalances, neurologic defects or are side effects of necessary medication regimes. In addition, implantation of penile prostheses in cases of organic impotence is an increasingly popular surgical procedure. These clinical approaches to sexual inadequacies, accepted by medicine since 1970, represent one horn of the dilemma.

The opposite horn pictures the medical profession firmly backed into a corner by cultural influences. For example, when hospital admissions occur, a significant portion of the routine medical history is the section on system review. A few questions are asked about the cardio-respiratory, the genito-urinary, and the gastro-intestinal systems. But in a preponderance of hospitals no questions are permitted or, if raised, answers are not recorded about human sexual functioning.

Physicians tend to forget that they are victims of cultural imposition first and of professional training a distant second.

In other words, it's all right to treat sexual inadequacies if these complaints are brought to the physician's attention, but it is not acceptable practice to ask the patient if he or she is contending with any form of sexual distress. When hospitals do not allow recording of complaints of sexual disorders, yet detail those of other natural functions, medicine has taken a position that is tantamount to denying that sex is a natural function. The hypocrisy of medicine's current dilemma is a serious embarrassment to all health care professionals.

The problem, of course, is how to alter this untenable position. The answer is continued research productivity and responsible reporting

E. E. Shelp (ed.), Sexuality and Medicine, Vol. I, ix–x.

of currently controversial material in a format acceptable to the medical community.

The study of human sexuality is still in its infancy. Perhaps the paucity of secure information is due more to the multifaceted dimensions of the subject than to inherent shortcomings of research protocols or limitations of investigative talent. Of course, attempts to interpret human sexual behavior still encounter cultural roadblocks acting to limit investigative inquiry and frequently leading to evaluation of research results from psychosocial rather than scientific points of view. Yet despite these cultural roadblocks, the many faceted subject of human sexuality has stimulated multidisciplined investigative approaches to the subject.

Volume I of *Sexuality and Medicine* is an excellent example of how to present material of sexual content to the health care field. It has been standard operating procedure to justify discussions of controversial sexual material by initially diverting reader interest into the safe ports of reproductive physiology or to detailed ethical considerations of sexual behavior. Such is the methodology employed in this text.

From a background of historical data, reader interest is directed to the basics of reproductive physiology and discussions of conception and contraception. There is no better entry to a broad consideration of sexual material than such hearth and home subjects.

The strength of the volume resolves around the material of ethical consideration and the discussions of sexual preference and transsexualism which are well-rounded and provocatively presented. Finally, the chapters on philosophic approaches to human sexual behavior tend to provide an additional dimension to ethical considerations. Indeed, the entire index is one of balance and counterbalance that should contribute toward unseating medicine from the horns of its dilemma.

Perhaps volumes of the stature of *Sexuality and Medicine* will encourage further excursions in responsible journalism on this subject.

Masters and Johnson Institute,
St. Louis, Missouri, U.S.A.

W. H. MASTERS
V. E. JOHNSON-MASTERS

PREFACE
CHANGING CONCEPTIONS OF HUMAN SEXUALITY

> On the Day of Judgement, a vast line of hopefuls queued up at the Pearly Gates hoping to be let in. Eventually St Peter arrived and the little chap at the front of the queue started a lengthy discussion with him. After a while the rest of the line started to grow restless and their anxious muttering grew in volume, but still the earnest discussion continued. Then just as it seemed that the dialogue would go on forever, there came a small cheer from the man at the front. Slowly it grew to an almighty roar of approval. Finally the message reached the man at the tail end of the queue. 'It's all right, masturbation doesn't count!'

This old story prompts the question: masturbation doesn't count as *what*? As a sin? as a sexual sin? as sexual? Few of the faithful would bet on their being admitted to heaven if churches taught that masturbation was a sexual sin, and few churches are that insensitive to the limitations of men and women. Their theologies can avoid a near-empty heaven either by denying that masturbation is a sin or by denying that it is sexual; in the latter case, the act is still wrong, but not as damning as a sexual sin. But denying that masturbation is sinful borders on asserting that it is a quite felicitous pastime. The churches, then, had better deny that masturbation is sexual. Suppose that sexual acts were defined as those acts that in virtue of human physiology could lead to reproduction. Then masturbation is not immoral sexuality, and not even perverted sexuality, for it is not a sexual act at all. Professor Goldman has written that "we all know what sex is, at least in obvious cases, and do not need philosophers to tell us" ([10], p. 122). *Au contraire*, as the joke shows. Philosophers have often argued that masturbation, 'obviously' a sexual act, is actually not a sexual act, for all sorts of curious reasons (see [9]); the irony is that Goldman's analysis of 'sexual activity' entails that masturbation is not a sexual act (see [10], pp. 16–18). Not only

E. E. Shelp (ed.), Sexuality and Medicine, Vol. I, xi–xxiv.

philosophers are unsure of its ontological status; the men's bathroom graffito, "If you shake it more than two times, you're playing with it," shows that many chaps wonder about their ordinary, unreflective judgments.

It seems to me that the person who says, in 1980, "we all know what sex is" and the person who uttered the same words in 1880 or 280 (in Latin) do not know the same thing. The concept of the sexual changes. What is accepted, done, believed, and discussed today in the area of sexual behavior is different from what was accepted, done, believed, and discussed in 1880 or 280. It is undeniable that changes in sexual behavior patterns and beliefs both contribute to and are influenced by changes in sexual concepts. One connection commonly mentioned is that occasioned by advances in birth control technology, but other connections are obscure and are best left to historians. We tend to think that modern conceptions of sexuality and contemporary sexual behavior are superior and enlightened behaviors and concepts. In particular, the 20th century is self-righteously proud of its sexual revolution, the rejection of what we take to be the restrictive and debilitating Victorian style of sexuality. But I am doubtful about these snap judgments of progress, and equally doubtful that all the changes occurring in the 20th century have amounted to anything fundamental. Elizabeth Hardwick expresses a similar uneasiness in her 'Domestic Manners':

> Sex, sex – what good does it do anyone to 'study' more and better orgasms, to open forbidden orifices, to experiment, to put himself into the satisfaction laboratory, the intesive care ward of 'fulfillment.' The body is a poor vessel for transcendence. . . Satiety, in life, is quick and inevitable. The return of anxiety, debts, bad luck, age, work, thought, interest in the passing scene, ambition, anger cannot be deferred by lovemaking. The consolations of sex are fixed and just what they have always been ([4], p. 6).

Sexuality and Medicine might, then, appear to be more of the same. But if the mechanics and the physiology of sexuality have been dissected and murdered, so much more reason is there to develop and continue the study of the ethical presuppositions and implications of what we are doing. Besides, within all her pessimism, Hardwick displays a bit too much optimism: it is just not the case that for many people satiety, let alone a smattering of satisfaction, is inevitable. It is true that Alex Comfort and Masters-Johnson have made us more aware of our sexuality and thereby more obsessed with sexual pleasure and performance, perceiving failure where there once was none. Nevertheless, there are many whose sexual dissatisfaction is only remotely the result of 20th century sexology, and unhappiness abounds. Reasons for this sorry state of affairs, reasons that go beyond fate, must be sought and reflected upon philosophically. *Sexuality and Medicine*, the first conscious, comprehensive effort to extend the megadiscipline of biomedical ethics to questions about human sexuality, is therefore a much-needed and desirable enterprise.

The concept of the sexual has changed and is likely to change again. Prior to recent times, sexual activity and reproduction were largely inseparable and the concept of the sexual reflected that fact: consider the words and phrases in the English language ('genitals,' 'sexual reproduction') that embody the connection. It is conceivable that in very early times, before humans realized that sexual activity caused pregnancy, the concept of the sexual did not yet include reproduction. In a sense the 20th century is attempting to return to those earliest times, and congratulates itself for finally achieving some separation of the pleasurable uses of the genitals and their procreative uses; the concept of the sexual accordingly is modified. Sexual *desire* is not exactly the desire to reproduce: one can desire intercourse and take precautions against fertilization, and one can desire to have offspring and prefer to accomplish it by artificial insemination. Sexual *activity* is not exactly activity that does or could lead to reproduction: fellatio and cunnilingus can be done for their own sake. Sexual *health* and illness are not exactly reproductive health and illness: homosexuals and the sterile can have satisfactory sexual lives yet never reproduce, and some who are physiologically able to reproduce may be too disturbed psychologically to engage in sex with another person. And sexual *pleasure* is not exactly reproductive pleasure: the pleasure of anal intercourse is not a reproductive pleasure, and the pleasures of birth are not sexual.

Such as they are, these are the hard-gained lessons of Anglo-American analytic philosophy mirroring the progressive consciousness of the 20th century. "It is obvious," writes Professor Goldman, "that the desire for sex is not necessarily a desire to reproduce . . ." ([10], p. 123). *A* and *B* are not the same thing, and are not to be understood in terms of each other, because *A* can exist in the absence of *B* and *B* in the absence of *A*. I am not convinced by this sort of logic. Imagine a world in which every human sexual act is accompanied by a desire for sexual pleasure and a desire to reproduce. Even in this world Goldman would insist that the concept of the sexual is still analyzable independently of the concept of the reproductive, because sexual desire and the desire to reproduce, only contingently correlated, *could* exist separately. But for the people in that imaginary world the concept of the sexual, their idea of what sexual activity is, would surely include reference to the reproductive, and not because they are sloppy thinkers. The logic of 'not necessarily' would sound hollow to them, given their practices, biological constitution, and especially their phenomenal experiences. One need not, in our world, take the heroic route, in order to preserve a conceptual connection between the sexual and the reproductive, of saying that the desire for sex *is* the desire to reproduce at some suitable level of the psyché ordinarily hidden from conscious inspection. Even though it might be unreasonable to claim that in our world fellatio and cunnilingus, for example, are not sexual acts precisely because they are not reproductive (or that when fellatio and cunnilingus are not part of a sequence of acts that culminates in a reproductive act, those instances of fellatio and cunnilingus are not sexual acts), we could easily adjust to this way of speaking, since our world is not altogether different from the imagined world. Less perverse is the following alternative way of retaining a milder connection between the sexual and the reproductive: sexual acts are those that produce the same pleasurable feelings normally produced by reproductive acts. There is a family resemblance that ties together the varieties of sexual activity and the central case of a reproductive act. Even though some sexual acts are not reproductive, this does not mean that the sexual has no conceptual interdependence with the reproductive. The similarity of the pleasure, despite contraception, guarantees that there is more than a mere causal connection between sexuality and reproduction.

Sara Ruddick has suggested that the analysis of sexual concepts by male philosophers is infected with bias, in particular the notion that sexuality is to be analyzed independently of reproduction because the former is a psychological concept and the latter is a physiological concept ([1], pp. 91–92). Just above I wrote that sexual pleasure is not reproductive pleasure, not in the sense that the pleasure of sexual acts is not the same pleasure as the pleasure of acts that normally lead to reproduction, but in the sense that the pleasures of sexual acts are not the same as the pleasures of giving birth. This claim is probably shortsighted, and if it is false there is reason to doubt that a gender-neutral analysis of sexual concepts is possible. Some years ago Alfred Kinsey speculated that women who dream about pregnancy consider the dreams to be sexy because these women *know* about the causal connection between sexual activity and pregnancy. It would seem to follow that women who dream about pregnancy, indeed women who daydream about becoming pregnant, should no longer consider these dreams and fantasies sexy because by now they know that contraception has severed the causal connection. (Even Kinsey's subjects had this knowledge, which makes his explanation implausible.) But many women still report that thinking about pregnancy is sexy. Could it be, as Geoffrey Gorer proposes in his critique of Kinsey ([3], pp. 175–176), that for women there is more than an intellectual connection between sexuality and reproduction, a relationship that is too-neatly ripped apart by analytic philosophy? That birth and its subsequent activities are fully sexual for women precisely in terms of the nature of the accompanying pleasures? Alice Rossi has no doubts about it:

> Whether the sucking [of the breast] is by an infant or a lover, oxytocin acts upon the basket cells around the alveoli, causing them to constrict. . . . The interconnection between sexuality and maternalism makes good evolutionary sense. By providing some erotogenic pleasure to the mother of a newborn baby, there is greater assurance that the child will be nursed. . . . Pregnancy and childbirth in turn improve the gratification women derive from coital orgasm. . . . Provide a woman with a rocking chair, and the far-away look of pleasure one often sees among nursing mothers is much closer to the sensual Eve than to the saintly Mary ([8], pp. 17, 29).

And, I might add, the father of the newborn baby can almost equally enjoy the sexual pleasures the infant offers; cuddling is sensually enjoyable to both the cuddler and the cuddled.

Now, I do not mean to buy into sociobiology hook, line, and sinker. Far from it. The point remains whether the connections for women between sexuality and maternalism are fully biological, fully cultural, or a mixture. That point is simply that in the analysis of sexual concepts the phenomenal experiences of those whose concepts they are and of those about which we are talking, must be given their due, regardless of the origin of those experiences. For an analytic philosopher to claim that we can analyze sexual concepts without mentioning reproduction, that we can understand what it is for an activity to be a sexual activity without bringing into the discussion the various aspects of reproduction, is to ignore what too many women (Eastern and Western, black and white, liberal and conservative) think, feel, and believe about their sexuality. It is not surprising, then, that when feminist philosophers analyze sexual concepts, they do not lose sight of reproduction. One of the strongest statements I have seen appears in a recent monumental work by Alison Jaggar:

> Limitations on women's reproductive freedom have been used to control their sexual freedom; for instance, men in early societies used a ban on birth control to force monogamy on women. Conversely, limitations on women's sexual freedom have been used to control their procreative freedom; most obviously, forced heterosexuality has also forced women into motherhood. . . . [S]exual freedom for women is not possible without procreative freedom and procreative freedom is not possible without sexual freedom ([5], p. 322).

In terms of analytic philosophy itself, if it is not possible to have *A* but not *B*, and if it is not possible to have *B* but not *A*, then *A* obtains if and only if *B* obtains. In the case at hand, Jaggar asserts that from women's point of view sexual freedom and reproductive freedom are inseparable; women have either both or (more likely) neither. Now, this tight connection for women between sexual *freedom* and reproductive *freedom* does not entail an equally tight connection between sexual activity and reproductive activity. Nevertheless, that some analytic philosophers have ripped apart the sexual and the reproductive by analyzing sexual activity, desire, and pleasure quite independently of the reproductive could misleadingly suggest, and probably has suggested, to others or themselves that there is no point in exploring any connections at all between the sexual and the reproductive, in particular between sexual freedom and reproductive freedom. It was left to the feminists to display to the world a lacuna in our knowledge created by the blindness of analytic philosophy to the experiences of women. The men had been busy analyzing freedom, to be sure, and sexuality, but did not think that the two analyses should proceed together or relate to reproduction. For whatever reason, men forget babies or unconsciously repress their knowledge of them when leisurely or professionally discoursing about sexuality and freedom. For men, sexual freedom is the freedom to have sex, full stop. (Love, too, we know from the same analytic philosophy, has nothing whatever to do, or 'not necessarily,' with the above items.)

I do not pretend that there is nothing left to be said about sexuality and reproduction; and there will be much to think about when the cloning of human beings and entirely artificial fertilization and gestation are commonplace. Contraception severed one link between sexuality and reproduction, but these other technologies will do more, allowing also reproduction without sexuality. In the meantime, there are other sexual concepts that require careful and unbiased attention, including, among others, homosexuality and rape. Both of these concepts have been undergoing change.

There is still widespread disagreement over what categories to employ in the definition of 'homosexuality,' in constructing criteria for the identification of the homosexual, and in the evaluation of homosexuality. These are not easy or idle questions. Sexual activity can be, and has been, conceived so narrowly that for an act to be sexual it must involve the insertion or penetration of something or other into an orifice of some sort of a person's body. As a result, women who engaged solely in cunnilingus with each other were not conceived of as homosexuals, and did not conceive of themselves as homosexuals, thereby avoiding both the moral and the medical evaluation of that behavior ([6], p. 9). Much kissing and hugging among women even today is explained away as the mere expression of female friendship or of the feminine propensity for affection. It is not conceived of as explicit sexual behavior because we refuse to recognize the sexuality of sensuality and because male culture prefers (for the sake of its own peace of mind) to overlook the threat of female homosexuality under its very nose. The male wisdom is that women are *very* sexual beings, and the men are right. But the men thought, in self-flattery, that women crave men. Not *nearly* necessarily. The sexuality of women is diffuse enough to gain appreciable satisfaction from babies and other women. The problem raised here in the understanding of homosexuality is that to the extent that we lack a sufficiently clear notion of sexual activity, we also lack a sufficiently clear notion of a homosexual sexual experience.

Another, by this time classical, problem is whether behavioral or mentalistic factors are to be referred to in the criteria for distinguishing heterosexuals from homosexuals. Suppose that in the style of Kinsey we employ a strictly behavioral scale according to which a heterosexual is a person who has had sexual experiences only with persons of the other sex and a homosexual is one who has had sexual experiences only with persons of the same sex. What do we make of persons who have had no sexual experiences at all? It would be strange to say that they were neither heterosexual nor homosexual. The complications are baroque for persons who have had both kinds of sexual experience:

> . . . the fact that women who are not lesbians greatly enjoy sex with other women was plainly evident at every party we went to ([11], p. 249).

On our behavioral scale, these women *are* 'partial' lesbians, and no one should feel compelled to deny it. At the same time, however, the notion of a 'partial' lesbian is quite mysterious. Consider a forty-year-old woman who had only heterosexual experiences at ages sixteen through twenty, and thereafter only homosexual experiences. Is she 'mostly' homosexual because twenty years is longer than five years? What if she had had heterosexual experiences with ten different males when she was younger, but only one continuous homosexual lover thereafter? Is she then 'mostly' heterosexual because the great majority of her partners had been persons of the other sex? We might try saying that through age twenty she was heterosexual, and thereafter homosexual, eliminating 'mostly' talk. But this tactic fails: chopping up the life of a person who has a mixture of sexual experiences means that he or she could be a heterosexual on Tuesday and a homosexual on Wednesday. Furthermore, the behavioral criterion apparently gives the wrong answer in the case of a person who has had only heterosexual (and perhaps unsatisfactory) sexual experiences, and fantasizes about or yearns for the homosexual experiences that are for practical reasons unattainable. A mentalistic criterion, which could take into account fantasies, wishes, desires, and the like, fares no better. Many sexually active heterosexuals have dreams and flashing fantasies of a homosexual nature, yet are only struck by their oddity, not being moved to action. It is difficult to distinguish those mental contents that do indicate homosexuality from those very similar mental contents that do not; in a sense, the problem of setting up the criteria for distinguishing between heterosexuals and homosexuals has only been pushed to a different level. Finally, mentalistic criteria of homosexuality are often (inescapably?) grounded in obscure, metaphorical, or untestable theories or pictures of the human mind. The lesson, I think, is that there is enough of a family resemblance between heterosexuality and homosexuality, in terms of the pleasures afforded by these activities, to make the search for distinguishing criteria futile and unnecessary.

It is in part for this reason – there are no firm *sexual* criteria for distinguishing between heterosexuality and homosexuality, and therefore no firm *sexual* criteria for the identification of the homosexual – and in part to offset the categorization of homosexuality as an illness, that there is an increasing tendency in the gay community to emphasize that homosexuality is a lifestyle rather than a matter merely of sexual object choice *or* preference (see Rice [10], pp. 257–258). If homosexuality is conceived of as a lifestyle, and the gay person is thereby not conceived narrowly as a person of a certain sexual nature, it does not make much difference whether the sexual component of homosexuality is characterized behaviorally or mentalistically. But at the same time, those who oppose anti-discrimination statutes that would protect gays in employment and housing argue that because homosexuality is a lifestyle, these statutes are conceptually confused. If there are no firm criteria for distinguishing the homosexual and the heterosexual, then "homosexual" does not denote a group in the way "woman" and "black" pick out distinct groups that are entitled to protection. And if homosexuality's being a lifestyle implies that one is a homosexual by choice – in the absence of identifying criteria one is able to announce that one is gay ('come out'), but one can also merely *proclaim* that one is gay, in the way one belongs to a political party by self-definition – then the argument against the statutes points out that protecting gays in this manner entails that *any* sufficiently large self-defined group would merit the same legal treatment (e.g., chessplayers). One troubling view that requires more thought is that members of groups that are not self-defined, such as women and blacks, and homosexuals *if* homosexuality is not a matter of choice, merit protection *qua* women, blacks and homosexuals; but members of groups that are self-defined, such as Lutherans and Republicans, and homosexuals *if* homosexuality is strictly a lifestyle, are not as a matter of principle entitled to protection. This view does not entail that the clauses mentioning political creed, religious beliefs, and sexual object preference should be deleted from anti-discrimination statutes. It does entail that whether or not such clauses are included in these statutes is a matter of political pressure and not a matter of philosophical rationale. If there is no philosophical principle by which we could distinguish self-defined groups that merit protection from those that do not, if there is no algorithm that separates 'trivial' from 'important' interests, then the self-defined groups that 'merit' protection are simply those that have mustered the political clout needed to get the statutes worded in their favor. If the chessplayers can pull it off, so be it. How we conceptualize homosexuality, then, has serious legal implications.

If homosexuality is today less frequently characterized as an illness, so too is rape hardly discussed as an individual pathology. Controversy about the concept of rape has been stimulated by women organizing around rape and by feminist philosophical and social scientific studies of rape. The law, we know now, has long defined rape in peculiar ways. Tennessee law in 1955, for example, could deal with a man who coerced a woman at knife point to submit to cunnilingus only by charging him with violating its sodomy ('crimes against nature') statute, and not by charging him with rape – no penetration of penis into vagina, no rape (see *Rose v. Locke*, 423 U.S. 48, 1975; in this case there was also the legal question whether the man could be charged, in the absence of penetration, under the sodomy statute itself). An obviously reasonable definition of rape is this: an act of rape occurs when a person is forced into a sexual act, or when a person is a participant in sexual activity but has not consented to participate. But philosophical and practical problems plague even this natural definition. Do we have a sufficiently clear notion of what a sexual act is, and therefore of what a coerced sexual act is? Can we effectively specify the acts signifying that consent is either present or absent, and can we precisely state the conditions in which a verbal agreement does not count as genuine consent? Does every kind of force or coercion make a sexual act rape? Satisfactory answers to these questions are tough to come by.[1] No wonder that some have concluded that the natural definition is inadequate, and that it is faulty because it assumes that rape is a sexual act. There are, it turns out, some good reasons for refusing to conceive of rape as a sexual activity, and to redefine it as an assault.

Because the 'atomic' physical actions involved in a rape and those involved in a consensual sexual act are, at some 'brute' level of description, very similar, it might seem counterintuitive to deny that rape is a sexual act. Indeed, on Goldman's view rape is a (nonperverted) sexual act quite because in virtue of its 'normal' form it has the ability to satisfy sexual desire. Robert Gray moves one step toward the feminist redefinition when he proposes that an act may be sexual from the point of view of one of the participants, but not from the point of view of the other participant. Thus, one and the same rape may be a sexual act for the rapist but not a sexual act for the coerced participant ([10], p. 162). Gray's philosophical point is that the experience of sexual pleasure is definitive of sexual acts (rather than, as in Goldman, a measure of their quality; [10], pp. 135–136), so that a woman who is raped, experiencing no sexual pleasure, has not from her perspective engaged in a sexual act. But the feminist redefinition goes beyond Gray. Gray's account leaves open the possibility that when or if a woman does experience (unwanted) sexual pleasure during the rape, the rape is thereby a sexual act, and this implication does not satisfy the demand that rape not be characterized at all as a sexual act. Furthermore, Goldman's implicit assumption that rape is an activity that tends to fulfill sexual desire (and is therefore a sexual act), and Gray's implicit assumption that a male rapist does experience specifically sexual pleasure in the act (and, therefore, for him the act is a sexual act), are both rejected by the feminist redefinition. Men do not, it is claimed, rape because the act promises to satisfy their sexual desires, and even if men do derive sexual pleasure from rape, sexual pleasure is not the important pleasure they derive and is not the motivating pleasure.

The case for the feminist redefinition begins by emphasizing the violence and brutality of many rapes, the battering of women that makes a rape much more like an assault than a sexual activity. The point is that the 'atomic' physical actions involved in a rape and those involved in a common assault are, at some 'brute' level of description, *more similar* than are the atomic actions of a rape and those of an ordinary sexual encounter. That the assault involved in a rape contemporaneously or subsequently includes battery with or upon the sexual organs does not amount to a significant difference. After all, we do not have separate concepts for assaults carried out with the fist and assaults carried out with the feet, so there is no reason to carve out special conceptual space for an assault carried out by the penis; and if we do not conceptually differentiate between assaults carried out on the face and those carried out on the chest, there is no reason to have a special category for assaults carried out on the genitals. Furthermore, rape and assault have in common another property that is not shared by ordinary sexual activity. Lack of consent is a feature of rape and of (most) assaults. Here the lack of consent in rape is not being used to define rape, but as an additional reason to assimilate rape to assault. Finally, feminist social science argues that rape is motivated by desires that are found in many assaults: the desire to humiliate, degrade, and dominate another person. The pleasures or satisfactions of rape are not sexual pleasures but derive from the expression of power over another individual. One interesting implication of the redefinition of rape is that it is wrong to conceive of the rape victim as a 'sex object.' Rape, along with beauty pageants, pornography, and advertising, is often claimed to be one of the ways women are sexually objectified. And it is often said of the rapist that in considering only his pleasure he uses the victim as an instrument. But if the redefinition of rape is correct, the rapist is more like a bully than an inconsiderate lover.

Perhaps in legal contexts the feminist redefinition of rape is convincing, but as a general account of the nature of rape it has two important weaknesses. First, the redefinition fails to take notice of a large difference between rape and assault, that there is attached to rape a kind of shame that never attaches to assault, a shame best explained by recognizing the sexual nature of rape (see [2]). The existence of shame among rape victims suggests that from their point of view they have not merely been assaulted but have been violated sexually, and that they conceive of rape as sexual. I do not mean, of course, to bless the shame, but simply to point out a fact that the proposed redefinition overlooks. Second, feminist social science has confused the question of the definition of an act with questions about its causal antecedents or about its precipitating motivations. Suppose, for example, that many assaults were motivated by, caused by, a desire to relieve the victim of his or her wallet, the ultimate goal being to acquire enough money for several meals. If we were to define an act by referring to its immediate motivation, then these assaults are not assaults at all, but thefts. And if we were to emphasize the ultimate motive behind the act – the desire to eat – then these assaults would have to be reconceived of as "preliminaries to the ingestion of food." If we distinguish, as we should, the definition of an act from its causal antecedents, then we can say very plainly that the theft was accomplished by an assault, that the assault was perpetrated to provide a meal. Similarly, referring to the motivation behind a rape in defining the act prevents us from saying something just as plain, that the rape, a sexual act, was employed by the rapist in order to humiliate, that the rapist chose a sexual means to humiliate his victim. But because the natural definition of rape has not been faring very well of late, it is clear that much more thought is required on this very important issue.

If the clinical approach to understanding and treating homosexuality is no longer in favor because homosexuality is no longer conceptualized as an individual pathology, then how are medicine and biomedical ethics relevant? If sexual concepts are always in the process of change, can these disciplines be assured of more than passing significance in sexual matters? First, it is unlikely, for example, that all homosexuality is the carefree, guiltless, happy choice of a lifestyle; indeed, it is unlikely that all *heterosexuality* is the carefree, guiltless, happy choice of a lifestyle. In both cases there is distress and debilitation that can be approached usefully by medicine, and biomedical ethics can contribute by helping to establish appropriate goals, guidelines, rationales, and conceptual boundaries. Second, even if rape is the result neither of an uncontrollable male sexual drive nor of a demented, innately criminal personality, medical approaches to understanding the rapist are not obviously irrelevant. If feminist social science is at least partially correct in claiming that the central motivation in rape is the desire to humiliate, biomedical ethics can be especially helpful, in sorting out in what sense such motivations are moral defects and in what sense they are social disturbances manifested through individual agents: in either case, medical and legal questions are left quite open and deserve attention. Third, the "Table of Contents" of *Sexuality and Medicine* reveals the wealth of topics that can be addressed systematically, issues arising as our history grinds on and not only generates changes in old sexual concepts but also creates new sexual concepts. I invite the reader to explore in this volume the frontiers of conceptual research in sexuality, and to discern for himself or herself the relevance of these analyses to medicine and biomedical ethics.

Department of Philosophy,
University of New Orleans,
New Orleans, LA 70148,
U.S.A.

ALAN SOBLE

NOTE

[1] Robin Morgan stumbles over these questions while unpacking the implications of her definition of rape, which is an elaborate variant of the natural definition: "rape exists any time sexual intercourse occurs when it has not been initiated by the woman out of her own genuine affection and desire"([7], p. 136). Morgan explains that on her definition a woman who agrees to have sex with her husband when she would prefer to watch a late-night movie has been raped (pp. 136–137). But is her agreement not genuine consent simply because she prefers to do something else? And even if she has been coerced by her nagging husband, is this the sort of coercion that makes the act rape? What is ironic is that Morgan, in her attempt to encourage us to take rape seriously, has actually trivialized rape. The worst we can say about a husband who asks for and gets sex from his wife during a movie is that he has been rude. Pope Paul VI, I think, was much nearer to the truth when he wrote in *Humanae Vitae*: "It is in fact justly observed that a conjugal act imposed upon one's partner without regard for his or her condition and lawful desires is not a true act of love, and therefore denies an exigency of right moral order in the relationships between husband and wife" ([1], p. 137).

BIBLIOGRAPHY

1. Baker, R. and Elliston, F. (eds.): 1975, *Philosophy and Sex*, 1st ed., Prometheus, Buffalo.
2. Foa, P.: 1977, 'What's Wrong with Rape', in M. Vetterling-Braggin, F. Elliston, and J. English (eds.), *Feminism and Philosophy*, Littlefield, Adams, Totowa, N.J., pp. 347–359.
3. Gorer, G.: 1966, *The Danger of Equality*, Weybright and Talley, New York.
4. Hardwick, E.: 1978, 'Domestic Manners', *Daedalus* (Winter), 1–11.
5. Jaggar, A.: 1984, *Feminist Politics and Human Nature*, Rowman and Allanheld, Totowa, N.J.
6. Jay, K.: 1983, 'School for Scandal', *The Women's Review of Books* **1**, #3, 9–10.
7. Morgan, R.: 1980, 'Theory and Practice: Pornography and Rape', in L. Lederer (ed.), *Take Back the Night*, Morrow, New York, pp. 134–140.
8. Rossi, A.: 1977, 'A Biosocial Perspective on Parenting', *Daedalus* **106**, #2, 1–31.
9. Soble, A.: 1980, 'Masturbation', *Pacific Philosophical Quarterly* **61**, #3, 233–244.
10. Soble, A. (ed.): 1980, *The Philosophy of Sex*, Rowman and Littlefield, Totowa, N.J.
11. Symons, D.: 1979, *The Evolution of Human Sexuality*, Oxford University Press, New York.

INTRODUCTION

Many of the vexing issues in the philosophy of medicine and in bioethics involve conceptual and normative understandings of human sexuality. Scholarly and public discussions about the morality of technological reproduction and homosexuality, for example, often proceed in a seemingly endless fashion without a critical examination of the conceptual foundations for the arguments that are presented. It appears that terms like 'natural', 'artificial', 'health', 'sick', 'desire', 'pleasure', 'masculine', and 'feminine', among others, when used in discussions about human sexuality warrant careful explication. This sort of analytical inquiry should contribute to a better understanding of the meanings of these terms and the concepts they represent. Further, with a better understanding, one would hope that their relevance to debated issues would be clearer and that progress could be made toward reaching consensus about disputed subjects.

This sort of inquiry is the focus of this collection. In short, the essays presented in this volume are intended to help bridge analytic philosophy and theories of medicine, including medical classification and warrants for medical intervention. Some subjects are discussed generally and others are more narrowly focused. In both instances, however, the contributors have attempted to demonstrate the importance of conceptualizations and empirical findings to understandings of human sexual function, identity, and expression. The analyses, perspectives, and insights provided by the philosophers and physicians contributing to this volume indicate that the goal of attaining a valid, objective understanding of almost any matter related to human sexuality will be difficult. The conceptual and value differences that underlie current debates are significant. As long as this remains the case, normative judgments about human sexuality and the relation of medicine to it surely will be contended. The companion volume [1] to this collection explores more fully the matter of moral judgment, human sexuality, and medicine.

The contributors to this volume explore certain concepts in order to

E. E. Shelp (ed.), Sexuality and Medicine, Vol. I, xxv–xxxii.

demonstrate how they are related to understandings of human sexuality and judgments of what is and is not a medically or morally licit use, function, or expression of human sexuality. In addition, these discussions illustrate how conceptual understandings influence justifications for interventions in or judgments about a person's sexuality by the medical sciences. The essays are presented in two sections. The first section contains six essays that address descriptively and critically sexual concepts, phenomena, and therapies. The second section of seven essays analyzes notions of healthy sexuality, 'sick' sexuality, sexual categories, and sexual dysfunction.

Robert Kolodny initiates the discussion of human sexuality. In a wide-ranging essay, he outlines contemporary medical and psychiatric thinking about 'healthy sexuality.' His focus is on sexual behavior rather than on the full range of components of sexuality. Kolodny illustrates the difficulty of stipulating criteria for judgments of healthy sexuality by examining psychiatric classifications related to sexual desire. In a section on issues involving sexual function, the author surveys views on premature ejaculation and female coital anorgasmia. He also considers views about the healthfulness of such practices as masturbation, oral-genital contact, incest, and homosexuality. Kolodny points out that even though there is widespread disagreement about what constitutes sexual health, some specific therapies are helpful in resolving certain patient complaints. Yet he cautions that though there is a developing measure of success in treating sexual dysfunctions, current understandings of sexual health still rest on uncertain foundations.

The analysis of healthy and unhealthy sexuality is extended by Frederick Suppe, who claims that many widely held medical and psychiatric perspectives on human sexuality are suspect. Because of this past, and present failures to adopt more stringent scientific standards for evidence, Suppe thinks that what is truly known about human sexuality is being discovered by disciplines other than medicine and psychiatry. He illustrates medicine's poorly informed views on sexual behavior by reviewing the classification of masturbation as a disease in the nineteenth and twentieth centuries. In agreement with John Duffy, Suppe sees medicine's oversight in the sexual arena as an effort to assert a morally influential role in society without adequate scientific knowledge as justification. He argues that until moral presuppositions give way to valid evidence, medicine and psychiatry will justly continue to experience an erosion of their authority in sexual matters.

What is known or believed by the medical sciences regarding the origins of sexual preference is surveyed by Stephen Levine. Levine argues that sexual identity has three components: gender identity, sexual orientation, and sexual intention. Gender identity involves seeing oneself as biologically male or female. It is formed early in childhood and is considered the psychological foundation of sexuality. Sexual orientation refers to the sex of people or mental images of people that attract and promote sexual arousal. It has subjective (psychological) and objective (behavioral) aspects. Sexual intention refers to what an individual actually desires to do with his or her sexual partner. Levine points out that the possible combinations of these three elements are numerous, and that many important questions regarding sexual identity remain unanswered. As an illustration of the complexity and uncertainty of theories of sexual identity Levine examines homosexuality, noting that it is the most thoroughly studied unconventional identity structure. He cautions, in conclusion, that etiologic theories are very intricate and constantly changing, so that one should not expect simple answers to puzzles of sexual identity structure.

One of the more complex and controversial topics in human sexuality is reviewed by Leslie Lothstein – transsexualism. Lothstein reviews theories of transsexualism in order to address critical diagnostic, treatment, and moral questions faced by transsexuals and society. The author critically assesses a variety of views, including psychological, psychobiological and psychoanalytic theories. He provides a brief statement of his own hypothesis on the etiology of transsexualism, and concludes that it represents a borderline psychological pathology that ought to be treated psychologically rather than surgically.

By now it should be clear that there is little consensus among researchers regarding healthy sexuality, sexual practices, sexual preferences, orientations, or identities. Vern Bullough's historical essay shows that intellectual interest in sexuality, disagreements about sexuality, and competing normative understandings are longstanding. In some form, according to Bullough, sex research and therapy have been practiced since the beginning of recorded history. He concisely provides an historical overview of this field from ancient to modern times. The oldest writings include views on impotency, contraception, fertility, and miscarriage as well as specific instructions to correct deficiencies. The first recognizably scientific studies of sexuality were conducted by Aristotle, whose work represents a combination of keen observation and popular superstition. Aristotle's influence, supplemented by Arabic astrological beliefs concerning sexuality, was felt throughout the Middle

Ages. Beginning in the eighteenth century, Tissot and others focused on masturbation and the psychological and physical ill-effects supposedly stemming from it. Bullough points out how the moral and emotional reaction of these researchers to their topic resulted in a general hostility toward sexuality.

Modern sexology began in the latter half of the nineteenth century with the work of Richard Von Krafft-Ebing and, somewhat later, Havelock Ellis. Ellis' observational method in sex research makes him the forerunner of contemporary efforts in the field. Bullough observes that a general lack of interest in sexual matters by most branches of medicine led, early in the twentieth century, to its being studied almost exclusively by psychologists and psychoanalysts. Thus, following Freud, sex researchers at this time came to view sexual problems largely as psychiatric illness.

The importance of the psychiatric model for sex research began to decline after World War II with Alfred Kinsey's behavioral studies and the physiological data garnered by William Masters and Virginia Johnson. There followed development of therapeutic techniques for specific common complaints. Bullough claims that these developments have established sex research as a legitimate scientific pursuit and sex therapy as an important specialty.

The first section of essays ends with a contribution by Fritz Beller who addresses selected topics from the vast field of reproductive biology and some of their ethical implications. The author provides descriptions of female and male sexual physiology and functions, processes of fertilization, and embryonic development. He also characterizes the main types of contraception and their possible side-effects, and provides a separate section on abortion. The principal causes and treatments of female and male infertility are discussed, and in a concluding section the author identifies and comments on moral issues associated with his biological topics.

The second section of essays provides a more philosophical examination of sexual concepts and their use in medicine. Exploring the concept of 'healthy sexuality,' Alan Soble notes that there exists considerable dispute both regarding the meaning of health and regarding which profession is the proper guardian of health. The author provides a conceptual critique of several accounts of healthy sexuality, demonstrating how medical views on this subject are often colored by their authors' value judgments concerning what is sexually legitimate. Soble's central focus is on Christopher Boorse's understanding of health, but also includes discussions of the views of Max Levin, Judd Marmor, Robert Gould, Masters and Johnson, Irving Bieber, Richard Green, Charles Socarides, H. T. Engelhardt, Jr., and Michael Ruse. Soble believes that these and similar views about health, sex, and the role of the professions claiming authority in matters of health are products of competition among nonmedical as well as medical criteria in conjunction with social forces. He concludes that sexuality is "whatever it is that is chosen by free people." Understood in this way, much of the medical profession's authority in matters of healthy sexuality vanishes, and the issue is shown to be fundamentally a political one involving the difficult-to-define concept of freedom.

Joseph Margolis next deals with the notion that there is some natural norm for health and sexual relations. He examines the work of Horatio Fabrega, as well as Henry Veatch's 'essentialist' view of normalcy and Christopher Boorse's species-typical account of human functioning as it relates to sexuality. He is mainly concerned with determining to what extent theories of physiological functioning can contribute to normative judgments about the sexual health of individuals, and in the course of his discussion shows that it is impossible clearly to segregate biological from sociobehavioral factors. Margolis concludes that medical conceptions of sexual health and disease are not free from the influence of normative concepts external to medicine, and that therefore sexual health as a normative concept reflects both medical and moral concerns.

Freud's understanding of sexual perversion is the topic of Jerome Neu's essay, which focuses primarily on Freud's multiple criteria for defining perversion. Neu criticizes Freud's heavy reliance on a biological or reproductive criterion of sexual pathology. According to this criterion, perversion represents a truncated sexual development in the individual. Neu specifically examines homosexuality and foot fetishism in the light of Freud's perversion criteria, as well as in the light of sociological and behaviorist specifications of perversion. He concludes that theories built on these criteria are inadequate, and that one must look beyond sexual activities and desires to the thoughts that lie behind them in order to achieve a proper understanding of normality and perversion.

Sandra Harding's essay focuses on what some people hold to be the opposite of perversion – the natural. She reviews recent scholarship showing that conceptions of what is natural or normal in behavioral differences between the sexes are largely cultural constructions. She identifies several obstacles to establishing biological bases for sex differences beyond a certain 'reproductive minimum,' and encourages medical scientists to view sexuality in a wider context including cultural and environmental as well as physiological differences between the sexes.

Robert Solomon explores what is frequently taken for granted and unexamined critically. Solomon looks at the varied meanings of heterosexuality. He distinguishes 'nature's purpose' for sex from human purposes for sex, and shows that the latter are quite diverse, having a wide range of symbolic and practical components. The author criticizes many views of sexual functioning as taking the form of moral and ideological pronouncements. He then considers four dimensions of heterosexual identity: biological sex, gender identity, sexual orientation, and sexual ethics. Each of these dimensions generates sexual 'purposes' or 'paradigms,' from which Solomon concludes that the meaning of heterosexuality is far more robust than what medicine has had to say about it. Accordingly, if medicine is to achieve a fuller understanding of sex in general and heterosex in particular, it must study poetry as well as physiology.

Bisexual behavior, according to Eli Coleman, is frequently described but inadequately understood. He discusses several impediments to achieving an adequate understanding of bisexuality, including definitional problems, difficulties of determining the extent of its actual existence, and personal problems encountered by people who identify themselves as bisexual. Among these latter, Coleman specifically discusses psychological adjustment, conformity to a lifestyle, identity development, and tools for assessing sexual orientation.

The second section, and the volume, concludes with an essay by Peter Breggin who argues for a connection between consistently good sexual experiences and romantic love. In his survey of differing viewpoints on the relation of sex and love, Breggin shows that a strong connection between the two can be traced back to the myths of ancient Greece and to stories in the Hebrew Bible. According to Breggin, these ancient notions provide an ideal of romantic love which is still influential today. Breggin's own theory of sexual love sees the physical as a vehicle for spiritual relationships between persons. He shows how the many qualities involved in love come together to contribute to an individual's freedom to relate to another. Thus, liberty and love are twin principles of a good and happy life. Sex emerges as an expression of all that is most satisfying in human relationships. Yet, acted out in isolation from higher human experiences of a spiritual nature, sex is a source of multiple complaints or dysfunctions. Breggin concludes that a successful sexual life is inseparable from a good life.

The wide-ranging content of these essays suggests that much work remains to be done if understandings of human sexuality are to become clearer and if pronouncements regarding human sexuality are to become better grounded. The complexities and uncertainties noted by the contributors ought not to be taken as an excuse to ignore an area of investigation seemingly so important to personal existence and to intellectual inquiry. The dialogue between philosophers and physicians contained in this volume perhaps can be an incentive or inspiration for other capable scholars to join in the quest for conceptual clarity about matters of human sexuality. As the contributors have shown, human sexuality is not subject only to the explanations of medicine or philosophy. Other disciplines have legitimate interests. Their insights, along with those of medicine and philosophy, equally ought to be critically evaluated. This expanded inquiry and dialogue might result in the participating disciplines being enriched by the conversation with others about concepts and phenomena of common concern. But as importantly, perhaps there will be a more robust appreciation of human sexuality, a better understanding of the advantages and limitations of medical views regarding it, and a clearer sense of where medicine is and is not the proper authority to evaluate or control human sexual function or expression.

The contributors to this volume generously have given of themselves. As editor, they should know of my appreciation for their commitment to this project and for their patience while the collection passed slowly through the process of publication. H. Tristram Engelhardt, Jr., and Stuart F. Spicker, general editors of the *Philosophy and Medicine Series*, again provided me with appropriate counsel and assistance in the design of the volume. Their contribution, separately and together, also is appreciated. Jay Jones, my research assistant, diligently helped to prepare the essays for the publisher. Susan M. Engelhardt proofread the manuscript, saving it from oversights and errors that otherwise would have been overlooked. Each of these individuals has had an essential part in putting this collection together. I thank them all for their work. Finally, I thank the staff at D. Reidel Publishing Company who have enabled me to bring this volume to completion.

Institute of Religion, and
Center for Ethics, Medicine, and Public Issues,
Baylor College of Medicine,
Houston, Texas, U.S.A.

EARL E. SHELP

BIBLIOGRAPHICAL REFERENCE

1. Shelp, Earl E. (ed.): 1987, *Sexuality and Medicine, II: Ethical Viewpoints in Transition*, D. Reidel Publishing Co., Dordrecht and Boston.

SECTION I

HUMAN SEXUALITY

ROBERT C. KOLODNY

MEDICAL AND PSYCHIATRIC PERSPECTIVES ON A 'HEALTHY SEXUALITY'

It should come as no surprise to readers of this volume that physicians generally consider themselves to be a source of considerable authority on the topic of health. Nevertheless, it is also remarkable that medical and psychiatric perspectives on just what constitutes a 'healthy sexuality' have shifted considerably over the past hundred years, indicating both a change in beliefs informed by advancing scientific knowledge and a reconsideration of matters of 'health' and 'normality' in light of changing cultural mores and practices. Despite such an evolution of medical and psychiatric attitudes toward sexual matters, it is important to recognize at the outset that there is no single perspective on which all theorists, researchers, and clinicians agree, so that dissenting opinion exists to a greater or lesser degree on almost every topic that will be touched on in this essay.

To illustrate this point, it can be noted that most physicians and psychiatrists have come to accept the fact that sexual feelings and the need for sexual gratification are common in persons with serious illnesses or physical disabilities, regardless of their physical capacity for sexual functioning. This is a relatively recent viewpoint that seems to have begun to gain professional acceptance in the mid-1970s; prior to that time, concern for the sexuality of such persons was infrequently voiced and was typically dismissed with pronouncements that either "nothing can be done to help" or that attention to such issues would only divert necessary energy from the more important task of treating the illness or disability itself. Indeed, physicians and psychiatrists alike often seemed to subscribe to the notion that the disabled or the chronically ill were asexual and in any event were so dependent and child-like that they needed to be protected from any sexual impulses they might have [1]. Apparently this latter viewpoint still exists among some physicians, since there are many rehabilitation programs and chronic disease hospitals that include little or no sexual counseling in their services to patients.

E. E. Shelp (ed.), Sexuality and Medicine, Vol. I, 3–16.

A parallel example can be found regarding geriatric sexuality. Prior to the important studies of Masters and Johnson [21, 22], it was generally felt by physicians and psychiatrists that advancing age typically brought about a marked reduction in sexual capacity and that sexual dysfunction was commonplace – even normative – in the geriatric population. Although numerous studies have now shown the fallaciousness of such an assumption, less than a decade ago a federal Medicare official refused to approve payment for sex therapy for impotence occurring in a 72-year-old man on the grounds that such a condition was normal at that age [20]. Indeed, even today there are frequent examples of middle-aged men who complain to their physicians of waning sexual interest or difficulty obtaining or maintaining erections who are told, in effect, "At your age, that's just what you can expect." While such misinformed medical advice appears to be decreasing in frequency, the point is that it still exists.

In this essay, I will attempt to outline the principal dimensions of contemporary medical and psychiatric thinking about a 'healthy sexuality.' Since space does not permit consideration of the history of the evolution of such ideas, interested readers are referred to several other sources to obtain this fascinating and informative background [2, 11, 30].

THE NATURE OF SEXUALITY

A point of semantics is necessary at the onset of this discussion. The term 'sexuality' is generally used today to refer to all aspects of being sexual, so that it does not refer only to sexual acts or sexual behavior, but encompasses psychological, biological, sociological, and cultural components. Despite this, the focus here will be principally on sexual behavior, with substantially less consideration of other aspects of sexuality.

A logical place to begin is to consider current medical and psychiatric opinion on the timing of the appearance of sexual feelings and behavior during the life cycle. Today there seems to be a remarkable degree of consensus that sexuality normally evidences itself from the time of birth on both in terms of operational sexual reflexes (erections occurring in male neonates; vaginal lubrication occurring in female neonates) and in terms of the occurrence of sexual feelings [3, 5, 24]. There is virtually no serious contention any longer that young children are asexual beings, with sexual impulses only arriving at the time of puberty. Thus, the somewhat revolutionary viewpoints of Moll, Hirshfeld, and Freud, expressed in the late 19th and early 20th centuries, seem to have been vindicated with the passage of time, although they were initially greeted with great derision.

At the other end of the life-cycle spectrum, it is also fair to say that a general medical consensus exists today that there is no predetermined age at which the capacity for sexual feelings or sexual functioning ceases, with some persons continuing an active, satisfying sex life in their 70s, 80s, and 90s. It should be noted, however, that many older persons have impairments to their sexual functioning because of illness (e.g., diabetes, neurological disorders, etc.) and others seem perfectly content to abstain from sexual activity for a variety of reasons. There is no current concept of geriatric sexuality that suggests that such voluntary abstention from sexual activity is unhealthy, although there is evidence that prolonged sexual inactivity in late adulthood leads to atrophic changes in the female genitals [18].

That sexuality is functionally present throughout the lifespan, both in its physiological and psychosocial dimensions, is further shown by documentation of reflex activity of the sex organs during sleep. Thus, periodic erections occur reflexively during sleep in healthy males, and in healthy females there are sleep-associated periodic episodes of vaginal lubrication [17, 24]. These findings, in combination with other discoveries of the nature of human sexuality of the past two decades, have led Masters and Johnson to espouse a view that has now come into general acceptance by physicians and psychiatrists – namely, that sex is a natural function. The elegant simplicity of this seemingly obvious fact once would have been a source of heated argument; today, however, it is recognized that sexual functioning (on a physiologic level) is quite parallel to other fundamental body processes such as respiration or digestion. Stated another way, no one teaches a man how to have erections or a woman how to have vaginal lubrication, just as no one teaches an infant how to breathe or digest its food.

In light of this relatively new perception of sex as a natural function, the entire question of when voluntary sexual abstinence is healthy and when it is not has been singularly neglected in medical and psychiatric literature. There is little evidence that prolonged sexual abstinence has deleterious health consequences, and in some regards, abstention from sexual activity may protect against the acquisition of certain disorders such as cervical cancer or sexually transmitted diseases [13]. On the other hand, most authorities believe that sexual desire is a normal, healthy component of human existence; thus, its absence (or suppression) may be a sign of psychological (or even physical) disturbances. This point has been particularly noted as improved diagnostic methods have allowed the identification of conditions such as prolactin-secreting pituitary tumors that severely suppress libido in men and also interfere with erectile and ejaculatory functioning, so that some cases that in the past would have been thought to be instances of sexual apathy without any organic contributant are now recognized as not at all psychogenic in origin.

The quandary is that it is at times very difficult to be certain of the volitional nature of sexual abstinence and extremely low (or nonexistent) levels of sexual desire. In the last seven or eight years, there has been increasing recognition of a diagnostic category called 'inhibited sexual desire' that seems to match a substantial portion of the population of adults seeking sex therapy. This has been operationally defined by the following criteria in DSM–III ([7], p. 278):

> Persistent and pervasive inhibition of sexual desire. The judgment of inhibition is made by the clinician's taking into account factors that affect sexual desire such as age, sex, health, intensity and frequency of sexual desire, and the context of the individual's life. In actual practice this diagnosis will rarely be made unless the lack of desire is a source of distress to either the individual or his or her partner.

Thus, a person who professes to have absolutely no interest in sexual activity, whose partner claims distress at this state of affairs, will be generally judged to be suffering from a disorder (or health impairment) called inhibited sexual desire, whereas the same person *without* a partner will be judged to be healthy, or at least free of this particular disorder. This seems to be taking the contextual interpretation of definitions of health to an extreme position, but even worse problems may be created by applying diagnostic criteria without regard to life circumstances.

There is also little agreement on what constitutes an 'unhealthy' amount of sexual desire at the opposite extreme. Certainly those persons who are so obsessed with sexual desire that they are prevented from functioning effectively in other aspects of their lives would be judged to be in a relative state of disability; especially when the dimensions of this obsessive preoccupation with sexuality is translated into acts that involve harm to others (e.g., rape, sexual assault, pedophilia) or behavior that breaks the law (e.g., obscene telephone calling, exhibitionism, voyeurism), there is general agreement in the medical and psychiatric communities that an unhealthy state exists. In the less extreme case, however – for instance, a fetishist whose propensity for collecting women's shoes and engaging in erotic rituals with them breaks no laws and harms no one – the line of demarcation is murkier, particularly if the individual is able to hold down a job and function effectively in interpersonal interactions. (Most medical and psychiatric authorities would hold that if this fetishist were unable to function sexually with another person without the use of women's shoes as erotic 'props,' or without at least fantasizing about the fetish object, his condition would not be entirely healthy, but there is some disagreement in the matter of fantasy content necessary for sexual arousal.)

Taking the discussion one step further, it should also be noted that people who have willing partners for very frequent sexual activity are not generally thought of as leading 'unhealthy' sex lives, although most authorities would probably be unwilling to call such a couple who engaged in sex dozens of times a day exemplars of a well-balanced life. Thus, the dilemma of quantifying 'excessive' frequencies for sexual activity remains very incompletely addressed in the medical and psychiatric literatures, perhaps in proper reaction to the intense scrutiny this topic received in the 18th and 19th centuries, when it was believed that loss of semen had disastrous effects on the brain and spinal cord and that it was equivalent to the loss of forty times its weight in blood [11].

Interestingly, DSM–III also includes in the residual category, Psychosexual Disorder Not Elsewhere Classified, a criterion for "distress about a pattern of repeated sexual conquests with a succession of individuals who exist only as things to be used (Don Juanism and nymphomania)"; with no other guidance provided as to what constitutes a 'pattern,' 'sexual conquests,' or a 'succession of individuals,' psychiatrists presumably must make this diagnosis on the basis of flimsy evidence indeed.

ISSUES INVOLVING SEXUAL FUNCTIONING

There is general agreement in the medical and psychiatric perspective that certain types of difficulties in sexual functioning are abnormal. For example, men who are unable to obtain or maintain erections in a substantial portion of sexual opportunities are seen as sexually dysfunctional, as are men who are unable to ejaculate intravaginally (as well as men who are unable to ejaculate at all). Similarly, women who are prevented from participating in pleasurable coitus because of either vaginismus (a condition of involuntary spasms and constriction of the muscles at the vaginal outlet) or because of persistent dyspareunia (painful intercourse) are also judged to be suffering from sexual dysfunction. These categorizations apply regardless of the etiology of the specific dysfunction – that is, whether the sexual problem results from purely organic causes or from purely psychosocial factors, or from a blend of both types of causation – and also pertain even in instances when there may be a reasonable psychological explanation that the dysfunction serves an ego-protective function. One caveat is in order here: in deciding when a sexual dysfunction is present as a sign of lack of sexual health, most physicians and psychiatrists would overlook isolated or infrequent episodes of the type mentioned above. That is, recognizing that for a variety of reasons (e.g., fatigue, inebriation, stress, a temporary state of poor health, preoccupation with something else, a grief reaction, etc.) people may have occasional lapses in their sexual responsivity that do not constitute a true syndrome of sexual dysfunction by their lack of persistence or relative infrequency. It is fair to say that absolute consistency of sexual response is no more the hallmark of sexual health than absolute consistency in memorizing strings of numbers is a prerequisite sign of healthy intellectual functioning.

There are several areas that are less agreed on by medical and psychiatric authorities regarding what constitutes healthy sexual functioning. The less problematic of these from the viewpoint of the clinician is deciding if a male who ejaculates rapidly falls into the category of 'premature ejaculation.' Although there have been various attempts over the years to define this condition precisely, the relevant point is that the diagnosis is not currently made unless the rapidity of the male's ejaculation is distressful either to him or to his partner. Thus, Kaplan suggests that premature ejaculation exists if the male doesn't have voluntary control over when he ejaculates [14] (a criterion that seems excessively stringent for what is, in fact, a reflex response), while LoPiccolo suggests a more pragmatic view – namely, that premature ejaculation *doesn't* exist if both partners "agree that the quality of their sexual encounters is not influenced by efforts to delay ejaculation" [19]. DSM–III couches its discussion of premature ejaculation in terms of 'reasonable voluntary control' and then goes on to say that "The judgment of 'reasonable control' is made by the clinician's taking into account factors that affect duration of the excitement phase, such as age, novelty of the sexual partner, and the frequency and duration of coitus" ([7], p. 280). Since in some other cultures a male's virility is judged by his speed in reaching ejaculation, and since as recently as at the time of the Kinsey reports it was claimed that over 75% of men ejaculated within two minutes of vaginal entry [16], the 'newer' view of greater control over the timing of ejaculation – which has had widespread cultural acceptance – may have less to do with health *per se* than with satisfactory relations between sexual partners and with the male's sexual self-esteem.

Far more controversial is the question of exactly what constitutes 'healthy' female sexual response. Some authorities, such as Masters and Johnson, believe that all women have the capacity for orgasm as a result of sexual stimulation. Others, such as Kaplan [14], suggest that a number of women lack the physiological capacity for experiencing orgasm during intercourse. In any event, most physicians would agree that few women experience coital orgasm during every coital episode and that substantially more women than men experience coital anorgasmia – never or rarely having orgasms during sexual intercourse. Thus, there is some question as to whether the condition of female coital anorgasmia is a normal condition (a view that certainly would have been popular in the Victorian era among physicians and psychiatrists, since most believed then that women were not capable of orgasm or sexual enjoyment) or whether it is in fact a matter of diminished sexual health. I believe that it is fair to say that most physicians and psychiatrists who have extensive experience dealing with sexual problems today would agree that female coital anorgasmia is, in fact, a sexual dysfunction, but it should be noted that DSM–III equivocates on this point:

> Some women are able to experience orgasm during noncoital clitoral stimulation, but are unable to experience it during coitus in the absence of manual clitoral stimulation. There is evidence to suggest that in some instances this represents a pathological inhibition that justifies this diagnosis [inhibited female orgasm; anorgasmia] whereas in other instances it represents a normal variation of the female sexual response [7].

Complicating this issue still further is the question of how to judge the frequency/consistency issue regarding female orgasmic response. That is, presuming that a woman wishes to be orgasmic, if she is orgasmic during only one-third of her pleasurable sexual opportunities, is she sexually dysfunctional or sexually healthy? Clearly, a man who failed to ejaculate or have orgasms (they are not, strictly speaking, precisely the same processes, although usually orgasm and ejaculation coincide in the male) in two-thirds of his sexual opportunities would be judged to be dysfunctional. Is it appropriate to apply the same quantitative standards on this issue to males and females, particularly since the woman's response may be somewhat more dependent on the male's staying power and proficiency as a lover than the male's response to his partner? What about the issue of orgasmic responsivity between two lesbian lovers? While these sorts of questions may generate heated discussions among physicians and psychiatrists today, there is certainly no uniformity as to the answers that might be given.

MISCELLANEOUS CONSIDERATIONS

Brief mention of several other aspects of sexual health deserve consideration. First, despite widespread agreement that masturbation is a normal, healthy form of sexual self-pleasuring, in the not-too-distant past medical and psychiatric opinion on this topic was quite different. In the 18th century, a Swiss physician – Tissot – promulgated the view that masturbation was a cause of nervous degeneration, insanity, epilepsy, and even early death [2, 11]. This belief persisted into the 20th century, and elaborate treatments were devised to 'cure' the problem, ranging from dietary restriction (since certain foods such as oysters, chocolate, ginger, and coffee were thought to irritate nerves supplying the genitals and thus provoke the impulse to masturbate) to a variety of mechanical devices such as metal mittens (to deter manual stimulation of the sex organs) and padlocked genital 'cages'. While we may smile today as we reflect on these quaint notions, we should recognize that it was only in the 1940 edition of Holt's *Diseases of Infancy and Childhood* that the discussion of masturbation was removed from a chapter on 'Functional and Nervous Disorders,' and that the use of surgery, mechanical restraints, or punishment to deal with this type of behavior was finally rejected.

Similarly, although Havelock Ellis and the Dutch gynecologist Theodore van de Velde wrote extensively about oral-genital sex as a healthy, pleasurable act in the early part of this century, there was considerable resistance to this notion in the medical and psychiatric communities until after the Kinsey studies [15, 16] appeared. Indeed, even today one can find physicians who are apparently so repulsed by the idea of oral-genital contact that they sternly warn patients to avoid this type of sexual contact.

A matter of much current interest and relative unanimity of medical and psychiatric opinion is that incest is an unhealthy form of sexual exploitation. Although there is clear condemnation of adult-child incest from almost all authorities, there is also some recognition that incest involving siblings, particularly those who are close in age, may not always represent an 'unhealthy' pattern and may simply be an outgrowth of natural childhood sexual curiosity. However, since recent data show that in a quarter of these experiences some type of force is involved, and since a quarter also involve siblings who are at least five years apart in age, the potential for exploitation and long-term psychological harm is not to be ignored [8, 26].

CURRENT PERSPECTIVES ON HOMOSEXUALITY

For the better part of the last century, the medical-psychiatric view of homosexuality has been strongly negative. It is unclear whether Freud, who wrote relatively little about homosexuality, was simply being compassionate or was stating his professional opinion in his much-quoted letter to the mother of a homosexual son [9] in which he noted:

> Homosexuality is assuredly no advantage, but it is nothing to be ashamed of, no vice, no degradation, it cannot be classified as an illness; we consider it to be a variation of the sexual development. Many highly respected individuals of ancient and modern times have been homosexuals, several of the greatest men among them (Plato, Michelangelo, Leonardo da Vinci, etc.). It is a great injustice to persecute homosexuality as a crime and cruelty, too.

Despite this uncertainty, the psychoanalytic community has been generally unified in seeing the homosexual orientation as a sign of neurosis and maladaptation. Sandor Rado, for example, explained homosexuality as a phobic response to members of the opposite sex [25], and psychoanalyst Charles Socarides (who wrote the chapter on homosexuality in the 2nd edition of the *American Handbook of Psychiatry*, published in 1974) characteristically labelled homosexuality "a dread dysfunction, malignant in character, which has risen to epidemic proportions" [28].

Despite the weight of this prevailing opinion of psychiatric orthodoxy, a new wave of research on the topic was initiated by psychologist Evelyn Hooker (who pioneered in selecting a sample of homosexual men who were neither prisoners nor psychiatric patients for her studies) and by psychiatrists Marcel Saghir and Eli Robins (whose studies matched non-patient samples of homosexuals and lesbians with non-patient samples of unmarried men and women). The findings that emerged indicated that homosexuality is not necessarily a form of psychological maladjustment, leading Saghir and Robins to state ([27], p. 317):

> Thus, it is quite inappropriate and scientifically untenable to label an individual psychiatrically ill because he happens to be a homosexual, for to do so would only tend to perpetuate the social and legal discriminatory practices against men and women who are primarily different in their sexual preferences but who otherwise show little other differences from their fellow non-homosexual men and women. Judgemental, moral and prejudicial stands are not constructive psychological approaches for dealing with any of the problems that a physician must face. . . . A condition cannot be made into a disease by simple intuition, moral indignation and proclamation.

As a result of such studies, combined with the awakening of what would eventually become the gay rights movement in America, intense pressures were brought to bear on the psychiatric community to abandon the 'disease' concept of homosexuality that had previously been in vogue. The details of this process, which are adroitly chronicled by Ronald Bayer [1], led to the eventual decision by the American Psychiatric Association in 1974 that homosexuality, *per se*, was not a mental illness, and to drop its inclusion in the 2nd edition of the *Diagnostic and Statistical Manual of Mental Disorders*. However, a new diagnostic category, called 'sexual orientation disturbance,' was disingenuously added to DSM–II [6]:

> This category is for individuals whose sexual interests are directed primarily towards people of the same sex and who are either disturbed by, in conflict with, or wish to change their sexual orientation. This diagnostic category is distinguished from homosexuality, which by itself does not necessarily constitute a psychiatric disorder.

This may have been perceived by the leadership of the American Psychiatric Association as a necessary political step to ensure passage of their decision to delete homosexuality itself as a diagnosable mental disorder; as it was, the vote was fairly close, with only 58% of the ballots cast favoring the decision [1].

In the decade since the American Psychiatric Association's fateful vote, there has been far more acceptance of homosexuality as a healthy variant of human sexual expression within the medical and psychiatric communities than there had been in the past. This acceptance can be seen, for example, in the appearance in well-known medical journals such as *Annals of Internal Medicine* and the *Journal of the American Medical Association* of articles describing appropriate history-taking and diagnostic methods with special applicability to the gay population. Furthermore, noted sex researchers Masters and Johnson published extensive data in 1979 showing that the physiology of sexual response among homosexual men and lesbian women is fundamentally identical to that of heterosexuals [23]. In addition, many gay physicians and medical students have come out of the closet, both personally and professionally, providing one other bit of evidence for the greater acceptance the medical community has extended toward homosexuality.

While there is little question that an evolution has occurred in medical and psychiatric attitudes toward gays, there is still some controversy over the issue of treating homosexuals who wish to convert to heterosexuality. Some psychiatrists, such as Thomas Szasz [29], indicate total opposition to the practice on the grounds that it is a strictly unwarranted, artificial approach to a problem that exists only because of social opprobrium; other psychiatrists are opposed because they do not accept claims that such therapy is likely to be successful [10]. These critics and others suggest that the appropriate treatment for such individuals should be geared at reducing unhappiness in their lives by making them more comfortable with their homosexual orientation [10, 31]. Nevertheless, a number of medical authorities, including Hatterer [12] and Masters and Johnson [23], hold to the opposite view, claiming that homosexual conversion therapy is frequently successful in their hands.

It should be noted that recently homophobic attitudes have surfaced again both in the general public and in the health care community as a result of the dreaded acquired immune deficiency syndrome (AIDS), which primarily affects homosexual males. While it is beyond the scope of this essay to examine this phenomenon in any detail, it should be noted that many health care workers have voiced major anxieties over contact with AIDS patients because of its extremely high fatality rate and lack of definitive knowledge of its precise mode of transmission. Shamefully, some hospitals and nursing homes have refused to accept patients with AIDS, and some physicians and psychiatrists seem to have reacted to the AIDS epidemic with an emotional sigh of relief, voicing the attitude that divine retribution is still operational. While it only seems to be a small minority of health care professionals who hold to such views at present, it is by no means certain that if the AIDS epidemic continues to grow for several more years, a more profound backlash may not become apparent in revising medical and psychiatric views of homosexuality.

A CONCLUDING NOTE

While there are many questions about defining just what 'healthy sexuality' means from the medical-psychiatric perspective, particularly as it involves quantitative issues, there now seems to be reasonable agreement that when distressing sexual problems exist, there is far greater amenability to today's treatment methods than was true in the past. Thus, sex therapy has enjoyed remarkably good success in short-term reversal of sexual dysfunctions, and non-psychotherapeutic techniques are also available for treating certain types of sexual problems – for instance, surgical insertion of a penile prosthesis can be used to treat organic forms of impotence, bromocryptine therapy reverses the hyperprolactinemia that sometimes causes inhibited sexual desire in men, and estrogen replacement therapy in the postmenopausal years can improve vaginal lubrication and reduce dyspareunia occurring as a result of estrogen depletion. Furthermore, recent attempts at using anti-androgens such as cyproterone acetate or medroxyprogesterone acetate in the treatment of chronic sex offenders also appear to offer some promise, although whether they can be said to offer a true chance at sexual 'health' is another matter entirely; perhaps at best we can hope that they will alleviate sexual suffering.

Despite these therapeutic advances, we must recognize the flimsy foundations on which we have built our current understanding of 'sexual health.' Until much more research work is done, we should readily admit the incomplete knowledge that we now possess, keeping open minds to new discoveries that lie ahead. For if the next quarter century sees anything close to the incremental increase in sexual knowledge we have gained in the last quarter century, we can be certain that our conclusions will be quite different from those we embrace today.

Behavioral Medicine Institute,
New Canaan, Connecticut,
U.S.A.

BIBLIOGRAPHY

1. Bayer, R.: 1981, *Homosexuality and American Psychiatry: The Politics of Diagnosis*, Basic Books, New York.
2. Bullough, V. and Bullough, B.: 1977, *Sin, Sickness, and Sanity*, New American Library, New York.

3. Calderone, M. and Ramey, J.: 1982, *Talking with Your Child about Sex*, Random House, New York.
4. Chipouras, S. *et al.* (eds.): 1979, *Who Cares? A Handbook on Sex Education and Counseling Services for Disabled People*, George Washington University, Washington, D.C.
5. Constantine, L. and Martinson, F. (eds.): 1981, *Children and Sex: New Findings, New Perspectives*, Little, Brown, Boston.
6. *Diagnostic and Statistical Manual of Mental Disorders*, 2nd ed., 1974, American Psychiatric Association, Washington, D.C.
7. *Diagnostic and Statistical Manual of Mental Disorders*, 3rd ed., 1980, American Psychiatric Association, Washington, D.C.
8. Finkelhor, D.: 1980, 'Sex Among Siblings: A Survey on Prevalence, Variety, and Effects', *Archives of Sexual Behavior* **9**, 171–194.
9. Freud, S.: 1935, 'Letter to an American Mother', reprinted in S. Arieti (ed.), 1959, *American Handbook of Psychiatry*, vol. 1, Basic Books, New York, pp. 606–607.
10. Gonsiorek, J. (ed.): 1982, *Homosexuality & Psychotherapy: A Practitioner's Handbook of Affirmative Models*, The Haworth Press, New York.
11. Haller, J. and Haller, R.: 1974, *The Physician and Sexuality in Victorian America*, Norton, New York.
12. Hatterer, L.: 1970, *Changing Homosexuality in the Male*, McGraw-Hill, New York.
13. Holmes, K., Mardh, P., Sparling, P., and Wiesner, P. (eds.): 1984, *Sexually Transmitted Diseases*, McGraw-Hill, New York.
14. Kaplan, H.: 1974, *The New Sex Therapy*, Brunner/Mazel, New York.
15. Kinsey, A. *et al.*: 1953, *Sexual Behavior in the Human Female*, Saunders, Philadelphia.
16. Kinsey, A., Pomeroy, W., and Martin, C.: 1948, *Sexual Behavior in the Human Male*, Saunders, Philadelphia.
17. Kolodny, R., Masters, W., and Johnson, V.: 1979, *Textbook of Sexual Medicine*, Little, Brown, Boston.
18. Leiblum, S. *et al.*: 1983, 'Vaginal Atrophy in the Postmenopausal Woman: The Importance of Sexual Activity and Hormones', *Journal of the American Medical Association* **249**, 2195–2198.
19. LoPiccolo, J.: 1977, 'Direct Treatment of Sexual Dysfunction in the Couple', in J. Money and H. Musaph (eds.), *Handbook of Sexology*, Elsevier/North-Holland, Amsterdam, pp. 1227–1244.
20. Masters, W.: 1977, personal communication.
21. Masters, W. and Johnson, V.: 1966, *Human Sexual Response*, Little, Brown, Boston.
22. Masters, W. and Johnson, V.: 1970, *Human Sexual Inadequacy*, Little, Brown, Boston.
23. Masters, W. and Johnson, V.: 1979, *Homosexuality in Perspective*, Little, Brown, Boston.
24. Masters, W., Johnson, V., and Kolodny, R.: 1985, *Human Sexuality*, 2nd ed., Little, Brown, Boston.
25. Rado, S.: 1962, *Psychoanalysis of Behavior*, Grune & Stratton, New York.
26. Renshaw, D.: 1983, *Incest – Understanding and Treatment*, Little, Brown, Boston.
27. Saghir, M. and Robins, E.: 1973, *Male and Female Homosexuality*, Williams & Wilkins, Baltimore.

28. Socarides, C.: 1970, 'Homosexuality and Medicine', *Journal of the American Medical Association* **212**, 1199–1202.
29. Szasz, T.: 1980, *Sex By Prescription*, Anchor Press/Doubleday, New York.
30. Tannahill, R.: 1980, *Sex in History*, Stein & Day, New York.
31. Woodman, N. and Lenna, H.: 1980, *Counseling with Gay Men and Women*, Jossey-Bass, San Francisco.

FREDERICK SUPPE

MEDICAL AND PSYCHIATRIC PERSPECTIVES ON HUMAN SEXUAL BEHAVIOR

> While the new [19th C.] scientific methodology led to discarding traditional theories, unfortunately not enough scientific knowledge existed to provide medicine with an alternative. In frustration and despair, the profession intensified its debates over medical theories and applied its traditional therapeutics – bleeding, purging, vomiting, sweating and blistering – even more drastically.
>
> John Duffy ([17], p. 4)

> What other medical condition requires that you wear rubber tits and dress up for a year as treatment?
>
> Pamela, a pre-operative transsexual

Other than somewhat improved anatomical and physiological understanding, in the areas of human sexuality medicine and psychiatry have not progressed much beyond the 19th century situation described by Duffy. Improved behavioral science research has overturned most traditional medico-psychiatric theories of sexual disorder (see e.g. [6, 7, 49]). Psychiatrists squabble over theories of sexual disorder in debates marked by a near total absence of scientifically credible non-anecdotal evidence. Bleeding, purging, sweating, and the like now take the form of catharsis on the therapist's couch, the bizarre cross-dressing therapies which, although psychologically debilitating [61], have become common in the treatment of gender dysphoria, etc. The inclusion of the paraphilias in DSM–III [3] is supported only by anecdotal or wishful evidence that fails to meet minimal scientific evidential standards [53, 54, 56]. The efficacy of psychotherapeutic techniques is unknown, and problems of control virtually preclude any objective empirical evaluation [26]. What would count as a cure is hopelessly obscure [52].

The history of 19th and early 20th century medical attitudes toward masturbation is a well documented episode of medical incompetence in which physicians served as moral arbitors providing 'medical support'

E. E. Shelp (ed.), Sexuality and Medicine, Vol. I, 17–37.

for conservative moral stances and reactionary political positions [9, 17, 18, 19, 20, 27]. Such moral compromises of medico-scientific objectivity have been frequent in the history of medicine – particularly as it concerns sexual practices and matters [9, 17]. Indeed, the central thesis of this paper is that the history of masturbation largely typifies medical and psychiatric perspectives on human sexuality even to the present, and thus that most of what these sister disciplines have to say about sexual behavior is as suspect as the deliverances of 19th century homeopathic medicine. I will attempt to defend this by a suitable marshalling of historical and other research. Further, I will suggest that insofar as we have any secure understanding of sexual behavior it comes from outside medicine and psychiatry, and that only by adopting the higher scientific standards of these other disciplines can psychiatry and medicine hope to regain its eroding authority over sexual matters.

MASTURBATION

In 1848, the superintendent of the lunatic asylum at Worcester published a sensational report to the Massachusetts legislature claiming 32% of admissions to the hospital were insane because of self-pollution (i.e., masturbation). The claim was taken up by medical pamphlet writers, quacks, and purity crusaders and quickly came to represent what "seemed to be the majority opinion in the medical profession at that time, or at least the majority opinion of doctors writing on the subject. Those few who remained skeptical of the relationship between self-abuse and insanity all too often became the pariahs of the profession" ([27], p. 203).

Thus the following sort of claims became commonplace in medical writings – well into the 20th century, as this 1909 account testifies:

SELF-POLLUTION

There are various names given to the unnatural and degrading vice of producing venereal excitement by the hand, or other means, generally resulting in a discharge of semen in the male and a corresponding emission in the female. Unfortunately, it is a vice by no means uncommon among the youth of both sexes, and is frequently continued into riper years.

Symptoms – The following are some of the symptoms of those who are addicted to the habit: Inclination to shun company or society; frequently being missed from the company of the family, or others with whom he or she is associated; becoming timid and bashful, and shunning the society of the opposite sex; the face is apt to be pale and often a bluish or purplish streak under the eyes, while the eyes themselves look dull and languid and the edges of the eyelids often become red and sore; the person can not look any one steadily in the face, but will drop the eyes or turn away from your gaze as if guilty of something mean.

The health soon becomes noticeably impaired; there will be general debility, a slowness of growth, weakness in the lower limbs, nervousness and unsteadiness of the hands, loss of memory, forgetfulness and inability to study or learn, a restless disposition, weak eyes and loss of sight, headache and inability to sleep, or wakefulness. Next come sore eyes, blindness, stupidity, consumption, spinal affection, emaciation, involuntary seminal emissions, loss of all energy or spirit, insanity and idiocy – the hopeless ruin of both body and mind. These latter results do not always follow. Yet they or some of them do often occur as the direct consequences of the pernicious habit.

The subject is an important one. Few, perhaps, ever think, or ever know, how many of the unfortunate inmates of our lunatic asylums have been sent there by this dreadful vice. Were the whole truth upon this subject known, it would alarm parents, as well as the guilty victims of the vice, more even than the dread of the cholera or small-pox ([62], p. 812).

Lest one not believe this account, a set of before (Plate 1) and after (Plate 2) photographs is offered, the after one being obviously and crudely retouched. Somewhat veiled versions of such dire warnings over self-abuse could be found as late as the 1950s in editions of the Boy Scout *Handbook for Boys* under a discussion of the law that "A Scout is clean."

John Duffy [17, 18, 20] argues that pre-Victorian American attitudes on sex were quite relaxed, documents a rising fear of sex beginning between 1830–34, and offers the following partial explanation:

What may have proved decisive in shaping the course of American thinking was the intervention of the medical profession. . . . Without an adequate basis to justify their position as a learned profession, they sought to emphasize their moral role in society. In consequence they happily seized on issues such as sexual morality and abortion, areas in which they claimed to have scientific knowledge, to bolster their standing in the community. By 1835 leading medical journals had picked up the themes of sexual excess and masturbation, and within a few more years the entire medical press was in full cry on the subject ([20], p. 7; cf. also [17]).

In addition, he cites the rapidly changing social order due to increased urbanization (p. 8). Related and compatible historical accounts are given by Bullough [9], Engelhardt [21], and Haller and Haller [27].

More generally, 19th century medicine asserted its moral authority not just over masturbation, but also abortion, contraception, venereal disease, and the inferiority of women and negroes. Thus diseases such as *Drapetomania* (the disease which causes slaves to run away) and *Dysaethesia Aethiopis* (hebetude of mind and obtuse sensibility of body – popularly called "rascality") which were peculiar to negroes emerged in

Plate 1

D. S. BURTON

The above is an illustration of D. S. Burton of Harris, Pa., before the habits of secret vice had begun to tell on him.

The illustration on the following page shows the same young man three years later taken when he had become an inveterate victim of the vice.

Plate 2

D. S. BURTON

The doctor's opinion was: "If this young man escapes the asylum he and his parents will be fortunate."

The instructions in this volume will save many a young man from swelling the list of the unfortunates that are in the asylums all over the country.

19th century southern U.S. medicine [21] and the suffrage movement was labeled as the rantings of women suffering from the psychosexual aberrancy of *Viraginity* (masculo-feminity) [59]. Over such 'medical issues' physicians generally supported the prevailing mores and attitudes of society, presenting 'research' that gave credence and medico-scientific legitimacy to such prejudices [17].

VICTORIAN AND POST-VICTORIAN SEXUAL ATTITUDES

The Victorian hysteria over masturbation transcends two quite different views on sexuality. Prior to 1880, chastity was the ideal among cultured Victorian males and females. "Marriage did not give licence to unreserved lust; the dominant middle-class ideal was chaste marriage, involving a moderate, restrained intercourse undertaken only for procreation. Although by 1880 that Victorian cult of pure, passionate, romantic love was being attacked by doctors it lasted in the United States, in middle-class culture, until about 1920" ([32], p. 141).

About 1880 medicine came to formulate a new "Theory of Normal Love" and began to develop the twin ideas of 'normal' and 'abnormal' love. "Lust was no longer ghettoized, isolated in a separate sphere of procreation or illicit pleasure, apart from love and passion. This new 'sex-love' was thought of as all-pervasive – as underlying, for example, religious feeling. Even the most 'spiritual' emotions were now thought of as touched by lust, as was all intense feeling. However, doctors now referred, not to a biblically named lust, but to a new, secularly named 'sexuality' to distinguish their formulation clearly from theologians" (*ibid.*, p. 141). The field of sexology was christened in 1904, and doctors began to view lack of erotic feeling and extreme continance as sexual aberrations (*ibid.*, cf. [23]). Closely allied was the emerging idea that emotions had genders or sexes, and by the 1880s and 1890s "the dominant medical meaning of the sexually 'normal' and 'abnormal' was the procreative and the non-procreative" where "the 'normal' and the 'abnormal' sexual were both alleged to arise from the physical and mental natures of males and females." "Late Victorian medical theorists of 'sexual perversion' pointed to normal, procreative male and female natures against which the perverse had turned. That procreative norm was presented in late Victorian medical literature as an innate, biologically determined imperative, a 'purpose' inherent in the structure and function of female or male bodies and organs" ([32], p. 142).

The Victorian prohibition against masturbation was, of course, compatible with earlier and later Victorian views. On the former it was to be condemned as a lapse in chastity. On the latter, it is one of the many forms of non-procreative sex. Indeed, in the 1880s masturbation quickly expanded to include all forms of non-procreative sex, including homosexuality. But because emotions had their genders – and effeminacy in males and butchness in females thus were unnatural – sexual perversion came to refer not only to non-procreative acts but also to failure to conform to sex-role stereotypes including what now are called transvestism and transsexualism. Thus by the end of the Victorian period, 'sexual perversion' lumps together masturbation, homosexuality, nonconformity to gender role stereotypes, transvestism, transsexualism, other non-procreative acts, and so on. Little attempt was made to differentiate between these conditions, classification terminology was not standard, and symptoms and consequences of one perversion were indiscriminately applied to any and all forms of sexual perversion. Remnants of earlier intersex views of Ulrichs and Hirshfield further confused matters.

The later Victorian Theory of Normal Love of course anticipates and lays a climate of acceptance for Freud's views on polymorphous sexuality and thus provides a natural avenue for psychoanalysis to become medicalized ([32], p. 153), despite Freud's belief that it was distinct from medicine. For the most part, during the 20th century medical and psychiatric perspectives so intertwined that attempting to distinguish medical from psychiatric perspectives becomes largely artificial and/or academic. In any case, they are dominated by the late Victorian "Theory of Normal Love" perspective most of this century.

While some early sexologists such as Havelock Ellis ([32], p. 148) attempt to distinguish forms of sexual perversion and provide diagnostic criteria, for the most part medical and psychiatric literature and practice tended indiscriminately to lump them together. For example, Dr. Charles L. Dana [14] lumps together masturbation, sexual murder, cannibalism (which he calls anthropophagy), flagellation, and exhibitionism as sexual perversions. Dr. James G. Kiernan [33] includes among the 'sexual perversions proper' psychical hermaphroditism, bisexuals (whom he terms heterosexuals), pure homosexuals, 'effemination or virginity,' gynandry and androgeny. In 1880 he treated a case of female cross-dressing as a case of nymphomania ([32], p. 196). Dr. Charles G. Chaddock's chapter on 'Sexual Crimes' [11] in a legal medicine text includes dress fetishism, sexual inversion, and all manifestations of the sexual instinct not in accord with propagation as perversions. And Dr. Allen McLane Hamilton in the same text [28] lumps together homosexuals, masochists, sadists, fetishists, mania hysterics, nymphomaniacs, and satyriasists together as perverts. These are not isolated examples, as the works surveyed by J. N. Katz [31, 32] indicate. Insofar as there was an attempt to distinguish types of perversion, it was on the basis of whether they were congenital, due to disease, mere vice, or on moral grounds. So long as surgical means such as castration (e.g. [6]) were preferred cures, classification of forms of sexual perversion was not terribly important. With the rise of psychoanalysis, there was a growing concern with aetiology and hence classification. Nevertheless, conditions such as homosexuality, transvestism, transsexualism, and effeminancy in men remain inadequately differentiated and confused until after World War II.

Indiscriminate lumping together of various sexual variations as sexual perversions is reflected in the fact that as late as December, 1962, the American edition of WHO's *International Classification of Disease* ([65], p. 70) only lists the completely undifferentiated category '326.2 Sexual Deviation.' Only in the 1967 eighth edition ([66], p. 148) do we find the differentiated entry:

> 302 *Sexual Deviation*
> Excludes: when associated with physical conditions (309).
> 302.0 *Homosexuality*
> Lesbianism Sodomy
> 302.1 *Fetishism*
> 302.2 *Paedophilia*
> 302.3 *Transvestitism*
> 302.4 *Exhibitionism*
> 302.8 *Other*
> Erotomania Nymphomania
> Masochism Sadism
> Narcissism Voyeurism
> Necrophilia
> 302.9 *Unspecified*
> Pathological sexuality NOS Sexual deviation NOS[1]

Similarly, the first 1952 edition of the American Psychiatric Association's *Diagnostic and Statistical Manual* [1] contains just the undifferentiated entry '52.2 Sexual deviation'; only in the 1968 second edition [2] do we find Homosexuality (302.0), Fetishism (302.1), Pedophilia (302.2), Transvestitism (302.3), Exhibitionism (302.4), Voyeurism (302.5), Sadism (302.6), Masochism (302.7), Other sexual deviation (302.8), and Unspecified sexual deviation (302.9) included as diagnostic subcategories.

Katz [31] and [32] reprints and/or abstracts a large and fairly representative sampling of 20th century medical and psychiatric writings on sexual deviation (with emphasis on homosexuality), and Lester [36] summarizes and evaluates additional literature. Analysis of these makes it clear that the late Victorian attitude of indiscriminately lumping together various variant sexual behaviors as undifferentiated sexual perversion was the norm the first half of this century and that transvestism, effeminancy, transsexualism, and male homosexuality were not clearly distinguished conditions until well after World War II. During this period, of course, masturbation continued to be viewed as undesirable by the medical profession.

Pre-World War II medico-psychiatric research on sexual perversion is most noteworthy for its inadequacy. Case studies, which at best provide raw data and establish nothing general, are the norm. When group studies are undertaken (e.g., [4, 29]), the subject populations are patient or criminal populations, adequate controls are not utilized, classifications are idiosyncratic, measures are inadequate, key notions and terms are poorly defined, statistics are crude by the then prevailing standards of the agricultural and biological sciences, and so on. Although diverse forms of perversion are labeled, individual therapists tend to give remarkably similar aetiological accounts for each of them. In short, very little in the way of reliable or objective findings emerges pertaining to sexual behavior, and published writings tend to 'confirm' the presuppositions of the researcher.[2]

To summarize, although there were some attempts to differentiate various forms of sexual perversion, as late as the beginning of World War II medicine and psychiatry typically employed a largely undifferentiated notion of sexual perversion that tended to conflate various standard patterns of variant sexual behavior discriminated by sexologists today. Medico-psychiatric research typically focused on discovering correlates or aetiology, and tended to offer much the same (e.g., pre-Oedipal conflict) explanations for diverse forms of sexual perversions as different as exhibitionism and homosexuality. Research evidence tended to be anecdotal, based on patient or criminal populations, and did not meet prevailing statistical standards for scientific research. In short, medicine and psychiatry were still operating under the late Victorian conception of normal and abnormal sex, lumping all unnatural sex together as sexual perversion or deviation, and tending to bolster its diagnoses and aetiological accounts with scientifically inept and worthless research. At the beginning of WWII Duffy's two 19th century characterizations quoted above are disturbingly apt.

WORLD WAR II AND ITS AFTERMATH

Up to WWII sexology seems to have been largely the province of medicine and psychiatry. During WWII, and especially afterwards, one finds a significant breakdown of that monopoly, with important research coming from outside the fields of psychiatry and medicine. The history of such changes has not been adequately researched, so the best I can do here is sketch some of the key developments and their plausible explanation.

In 1938 Alfred Kinsey, an entymologist at Indiana University who specialized in the gall wasp, began to study human sexuality via the interview procedure. He brought the entymologist's concern with classification to his work, was more able to view humans as just another species, and so was less inclined to pass moral judgment on human nature. Further, he had a sophisticated understanding of research design and statistics, something lacking in medico-psychiatric studies. His subjects were almost entirely non-patient, non-criminal in source – the importance of which Havelock Ellis [21] first had noted in 1895, but which sexologists largely had ignored. The publication of Kinsey's male and female volumes in 1948 and 1953 [34, 35] displayed how far removed typical American sexual practices were from the prevailing social mores and moral attitudes. Such clash of values and practices ultimately was resolved by society's altering the mores and moral views to conform more closely to practice. Doing so, of course, undercut the moral authority of defenders of the late Victorian view and the Traditional Western Sexual Norm ("ejaculation is to be confined to the vagina of one's spouse in a lifelong heterosexual marriage") – namely, medicine, psychiatry, and the clergy.

The scandal and hysteria surrounding the initial publication of the Kinsey reports led to attempts to challenge their scientific adequacy – an adequacy largely vindicated by a study commission of the American Statistical Society [12]. The holding of Kinsey's work to a scientific standard not previously imposed on medico-psychiatric sex research had the long-term effect of raising the scientific standards by which sex research generally was to be judged. Given the mediocrity of earlier work, coupled with the manifest failure of later medico-psychiatric attempts such as Bieber *et al.* [8], which tried to meet this standard [49], this helped to discredit the scientific basis on which medicine and psychiatry claimed their authority over sexual matters. Medicine and psychiatry's complicity in the McCarthy period's persecution of homosexuals may have further contributed to this decline in moral authority.

For the first time, during WWII the U.S. military attempted to use psychological tests to screen out undesirables, among them homosexuals. This led to the attempt to develop objective psychometric instruments that would diagnose homosexuality and other so-called perversions – which in turn required improved conceptualization of sexual perversions and an attempt to identify their correlates. During the 1940s and 50s also came the development of the MMPI, which attempted to develop a self-report instrument for the differential diagnosis of mental disorders. Difficulties with developing such a test for heterosexuality vs. homosexuality (the original conception of the *Mf* scale) [49] contributed to the clear differentiation of homosexuality from gender stereotype nonconformity (e.g., effeminism in males), transvestitism, and transsexualism.

In the 1950s Evelyn Hooker, a psychologist, began her research on non-clinical homosexuals – which simultaneously challenged the ability of standard Rorschach projective tests (e.g., the Wheeler signs) to diagnose homosexuality and also claims that homosexuality inevitably constitutes psychopathology. When gays began to lobby the American Psychiatric Association for removal of homosexuality *per se* from its catalogue of mental disorders [2], Hooker's research was offered as refuting evidence. Against it psychoanalysis offered studies such as the seriously flawed Bieber *et al.* one [8]. The battle over declassification erupted into the popular press in ways that tended to discredit the scientific status of psychiatric diagnosis and disease classification [5, 50]. Ultimately, the homosexuality normality view prevailed, tipping the research prestige from the medico-psychiatric practitioners to the non-medical, non-psychiatric social scientists such as Hooker, whose research supported the declassification.

This outcome was strongly conditioned and intensified by the formation of the National Institute of Mental Health Task Force on Homosexuality in 1967. The task force membership was a 'stacked deck' in favor of non-physicians and non-psychiatrists. Of its fifteen members, only three were M.D.s, and even these included advocates of progressive medico-psychiatric views on homosexuality such as Judd Marmor. Evelyn Hooker was Chairman of the task force. Why this task force was formed, and why it had such a stacked membership has not been historically documented despite the significance of its work. In any case, this commission, the large amount of research it sponsored, which generally was unsympathetic to, and unsupportive of, the prevailing Victorian medico-psychiatric orthodoxies, and its final report [37] were major forces in further undercutting medico-psychiatric authority over sexual matters, and hastened the rejection of standard medico-psychiatric views on homosexuality. Further replicative studies [6] and [7] by the Kinsey Institute later intensified these effects.

The Gay Liberation movement's challenge of the medico-psychiatric orthodoxy and its moral authority has been documented by Bayer [5] and D'Emilio [15]. This attack was an important part of the political solidification of the Gay Liberation movement and of the emergence of homosexuals as a distinctive minority in U.S. society [15]. As more and more homosexuals came out of the closet, openly gay scientists began to emerge. And some of those whose research expertises were appropriate to studying homosexuality began to do so. The *Journal of Homosexuality* was founded by openly gay researchers, and the bulk of its editorial board is openly gay. On its pages and in other standard journals, gay scientists began to do research whose designs were informed by their experiences as gays, their perceptions of what the relevant research issues are, and so on. And frequently their work meets the highest methodological standards. Research on homosexuality increasingly is dominated by gay researchers. Old problems such as aetiology have become passé, being replaced by new foci such as homophobia and 'coming out' phenomena. Revisionist science is openly advocated and practiced by gay scientists under the imprimatur of an arm of the American Psychological Association [45, 51]. So far as the 'perversion' of homosexuality is concerned, medicine and psychiatry have lost their moral or scientific authority, save for the obligatory appearance of living Victorian psychiatrists such as Irving Bieber or Charles Socarides in media special reports.

Medicine and psychiatry continue to maintain their authority over other forms of sexual perversion (the paraphilias of DSM–III), gender identity disorders, and psychosexual dysfunctions such as inhibited sexual desire, frigidity, and impotence. Elsewhere [54] I argue that (1) there is no competent empirical evidence justifying the inclusion of any of the paraphilias as *per se* mental disorders, and (2) precisely the same grounds that were used to justify the removal of homosexuality *per se* from DSM–II and III equally require the removal of the other paraphilias from DSM. Their continued inclusion is a vestige of moral, not scientific, authority over sexuality and constitutes unwarranted perpetuation of the late Victorian "Theory of Normal Love." Emboldened by the successes of Gay Liberation there are emerging liberation movements for the other perversions, including sado-masochism and paedophilia. Challenges to their medico-moral condemnation are being produced (e.g., [10], Ch 8, [26, 42, 57, 58, 60]). Studies on non-clinical, non-criminal populations (e.g., [46, 48]) are beginning to challenge the psychopathic orthodoxies of medicine and psychiatry. In short, there is every reason to believe that medicine and psychiatry's claimed authority over the paraphilias is as shaky and suspect as it was over homosexuality.

The medico-psychiatric establishment's control over gender identity problems, especially transsexualism, is on firmer ground. This is not because research in the transsexualism area is any better than it is for other Victorian perversions, but rather because the standard treatment is hormonal therapy and sexual reassignment surgery. Since these are subject to licensing requirements of the medical monopoly, it is unlikely that authority over these will be surrendered. Thus it is important to know how suspect current transsexual orthodoxies are.

DSM–III diagnoses transsexualism (302.5, also known as gender dysphoria) as:

A. Sense of discomfort and inappropriateness about one's anatomic sex.
B. Wish to be rid of one's own genitals and live as a member of the other sex.
C. The disturbance has been continuous (not limited to periods of stress) for at least two years.
D. Absence of physical intersex or genetic abnormality
E. Not due to another mental disorder, such as Schizophrenia ([3], pp. 263–64).

Criteria D and E reflect failure to find correlations with transsexualism, and C is included to avoid confusion of it with short-term disorders not requiring such radical treatment as castration and plastic surgery. The key criteria are A and B, which are essentially Christine Jorgensen's characterization of her condition in the *American Weekly* Sunday supplement to Hearst papers [30] minus her surgeon's aetiological speculations. (Note that DSM–III as a matter of policy is neutral with respect to aetiology in its diagnostic criteria.) This, coupled with the fact that persons presenting themselves for sex reassignment surgery typically catechismally present essentially the same symptoms ([40], p. 252) makes the diagnosis suspect. Commenting on transsexualism, Prince writes:

> I am not referring to the verbally expressed motivations such as 'I hate my penis'; 'I am a woman trapped in a man's body'; 'I think like a woman'; 'I have to have hormones and surgery, or I'll commit suicide.' We have all heard these statements *ad nauseum*. They are the catechism of those seeking surgery and their very identity from patient to patient ought to be a caution light to any professional working in this area. Each person is an individual and when a number of people say exactly the same thing for the same purpose, it is a fair assumption that the expressions have been gleaned from something written or said by another who was seeking the same solution to the same problem. The statements are made on the theory that if they worked before for someone else they might likely work again for the speaker ([44], pp. 263–264).

If there is a genuine transsexualism category, it almost certainly is not that of DSM–III.

Ostensibly because patients requesting sex reassignment surgery do not always fit the classic diagnostic conditions and an (insignificant) number of post-operative patients have subsequently undergone psychosis or had other psychological problems (e.g., [13, 38]), it is standard for gender-identity clinics to screen prospective patients harshly. A typical screening requirement is that the patient live and dress in the desired gender role for at least one year while receiving hormone therapy, electrolysis, and counselling. "The rationale behind this approach appears to be based on two beliefs: (1) that the candidate should try the role on 'for size' as completely as possible to see if it is what she really wants; and (2) that anything short of surgery is essentially reversible, and the candidate can 'back out' without any major problems if it appears the sex reassignment is inappropriate" ([41], p. 275). However, the second assumption is erroneous, since the use of hormones such as Premarin and electrolysis effects are not reversible, the effects of working as the opposite sex on the job often do irreversible harm to one's career, and the year-long passing period produces intense psychological stresses far greater than the stresses due to the gender-role nonconformity that typically occasioned the request for sex reassignment surgery in the first place (*ibid.*, cf. [61]). In short, evidence justifying this often cruel and barbaric treatment practice – which frequently proves as traumatic as Victorian infibulation and other cures for masturbation – is meager and inadequate. Indeed, based on reports from preoperative transsexuals its actual justification seems to be rooted in physicians' insecurity in their own masculinity and the persistent belief that males are superior to females (hence one ought not to make it easy for males to become inferior females, and there is no way any woman surgically can become a *real* man). That is, the screening device seems designed as a means for withholding surgical reassignment from all but the most desperate.

The research surrounding transsexualism is nearly as meager, inferior, and suspect as that surrounding the paraphilias other than homosexuality. And just as one finds Lionel Ovesey attempting to distinguish pseudohomosexuality from genuine homosexuality by fiat [39], in carbon copy form one finds his distinction between primary and secondary transsexualism at the center of debates over who should receive reassignment surgery [45]. Far more important than research in determining present-day conceptualization and treatment of transsexualism is the Victorian "Theory of Normal Love", which *inter alia* ascribed different natures to males and females and viewed deviation from these gender-role stereotypes as abnormal sexuality, hence a form of perversion. This Victorian view provides the only plausible justification for viewing transsexualism as a sexual mental disorder. Divorced from these Victorian underpinnings, the request for sex reassignment surgery is nothing other than a request for cosmetic surgery and medical aid in convincing the legal and other bureaucracies to allow one to change one's gender designation on various documents and enjoy the legal rights of one's new cosmetic sex. Although patient demands for gender reassignment typically are obsessive, sex reassignment surgery is no more the treatment of a medical disorder than a nose job is treatment for the medical condition of Judaism. As with other forms of vanity plastic surgery, it is possible for medicine to retain its authority over sex reassignment without indulging in its currently suspect scientific posturings.

Psychosexual distinctions such as inhibited sexual desire, frigidity, and impotence *per se* are medical problems only against a normalcy standard such as that of the late Victorians (but not that of the more romantic early Victorians), which demanded sexual desires and consummation. Absent such a medical standard, it is not *per se* a medical or psychiatric problem. For those wanting sexual desire or consummation, their absense is an ego-dystonic problem appropriate for medico-psychiatric treatment.

In summary, the weight of scientific evidence does not lend much support to medicine's and psychiatry's claims to authority in sexual matters. Traditionally the sexual views of medicine and psychiatry have been conceptually confused and rooted in Victorian moralistic views rather than based on credible scientific evidence. Medicine's and psychiatry's authority in the area of homosexuality has been largely undercut; and it is eroding with respect to other traditional perversions. Their hold on authority is secure only in those areas where standard treatments involve chemotherapy or surgery, which are licensed exclusively to physicians and/or psychiatrists. While such authority may be secured, its legitimacy is compromised to the extent medicine and psychiatry smothers it in scientifically illegitimate views (as tends to be the case with transsexualism). To the extent that medicine and psychiatry can claim legitimate scientific authority in sexual matters, it is only by virtue of rejecting traditional medical and psychiatric views, practices, and standards with respect to sexuality, and instead functioning credibly as members of the larger behavioral science community of scientists.

IF NOT THE MEDICAL MODEL – THEN WHAT?

The late Victorian "Theory of Normal Love" is, of course, a version of the 'medical model' in which disease or disorder is conceptualized as deviation from the normal. Not only does the medical model involve inventing the archetypally normal person, but through the nosological classification of abnormality it reifies deviation into archetypal patients. In the sexual realm this leads to what Foucault [24] calls the *invention* of The Homosexual, where in the 19th century medicine and psychiatry converted homosexuality from a component in an individual's psychosexual and behavorial repertoire into a species of individual:

> The nineteenth-century homosexual becomes a personage, a past, a case-history, and a childhood, in addition to being a type of life, a life form, and a morphology, with an indiscreet anatomy and possibly a mysterious physiology. Nothing that went into his total composition was unaffected by his sexuality. . . . It was consubstantial with him, less as a habitual sin than as a singular nature. We must not forget that the psychological, psychiatric, medical category of homosexuality was constituted from the moment it was characterized. . . . The sodomite had been a temporary aberration: the homosexual was now a species ([24], p. 43).

And just as psychiatry and medicine invented The Homosexual, so too they have invented The Transvestite, the Pedophiliac, The Sadist, The Masochist, The Voyeur *et al.*

The weight of post-World War II sex research not only calls for the deinvention of The Homosexual, The Transsexual *et al.*, but it also stresses the normally broad diversity found in psychosexual make-ups. Today researchers recognize a person's *sexual identity* as involving the following distinct components: biological sex, gender identity, social sex role, and sexual orientation comprising sexual behavior, patterns of interpersonal affection, erotic fantasy structure, arousal cue-response pattern, and self-labeling. (These are discussed in my [54]; see also [47, 52, 53, 55].) These components enjoy varying degrees of plasticity and stability at various ages, and provide such a rich fabric to one's sexual identity that one's sexuality is quite idiosyncratic. Being different, deviation from total conformity to some Victorian or any other sexual ideal is not abnormal; it is the very essence of normal sexuality. Being heterosexual is just as problematic as being homosexual or fetishistic. There is every reason to believe that essentially the same mechanisms are involved in producing each of these. Thus, what the invention of The Homosexual and The Paraphilias does is violence to the rich diversity of human sexuality and thereby inhibits its productive expression and incorporation into rewarding and valuable lifestyles. In stressing this I ultimately am drawing attention to something Kinsey knew and so perceptively expressed:

> Viewed objectively, human sexual behavior, in spite of its diversity, is more easily comprehended than most people, even scientists, previously have realized. The six types of sexual activity, masturbation, spontaneous nocturnal emissions, petting, heterosexual intercourse, homosexual contacts, and animal contacts, may seem to fall into categories that are as far apart as right and wrong, licit and illicit, normal and abnormal, acceptable and unacceptable in our social organization. *In actuality, they all prove to originate in the relatively simple mechanisms which provide for erotic response when there are sufficient physical or psychic stimuli.*
>
> To each individual, the significance of any particular type of sexual activity depends very largely upon his previous experience. Ultimately, certain activities may seem to him to be the only things that have value, that are right, that are socially acceptable; and all departures from his own particular patterns may seem to him to be enormous abnormalities. . . . There is little evidence of the existence of such a thing as innate perversity, even among those individuals whose sexual activities society has been least inclined to accept. There is an abundance of evidence that most human sexual activities would become comprehensible to most individuals if they could know the background of each other individual's behavior ([34], p. 678; italics added).

If medicine and psychiatry are to regain their lost authority over sexuality, they had best recognize the truth of Kinsey's insight, put to final rest the Victorian "Theory of Normal Love", and commit themselves to practice rooted in and constrained by scientific research meeting the highest contemporary standards. Such a medicine and psychiatry of sexuality will be quite unlike what we are accustomed to. But what authority they enjoy will then become as legitimate as their prior authority has been unscientific and fraudulent.

University of Maryland,
College Park, Maryland,
U.S.A.

ACKNOWLEDGEMENT

I thank John Duffy for advice and comments on the draft of this article.

NOTES

[1] Curiously, the 1948 edition [63] of ICD–6 lists as examples (much as Lesbianism and Sodomy are listed under 302.0 above) for the undifferentiated sexual deviation category: exhibitionism, fetishism, homosexuality, pathologic sexuality, sadism, sexual deviation. This practice is not followed in the later [64] or [65]. Mental disorders were first included in the 1948 edition of ICD [63]. The current edition is [67].

[2] The above generalizations are based on an examination of over 3000 journal articles and books on human sexuality published over the last century. Space limitations preclude listing them.

BIBLIOGRAPHY

1. American Psychiatric Association: 1952, *Diagnostic and Statistical Manual: Mental Disorders*, American Psychiatric Association, Washington, D.C.
2. American Psychiatric Association: 1968, *Diagnostic and Statistical Manual of Mental Disorders*, 2nd ed., American Psychiatric Association, Washington, D.C.
3. American Psychiatric Association: 1980, *Diagnostic and Statistical Manual of Mental Disorders*, 3rd ed., American Psychiatric Association, Washington, D.C.
4. Barahal, H. S.: 1939, 'Constitutional Factors in Psychotic Male Homosexuals', *Psychiatric Quarterly* **13**, 391–400.
5. Bayer, R.: 1981, *Homosexuality and American Psychiatry: The Politics of Diagnosis*, Basic Books, New York.
6. Bell, A. and Weinberg, M.: 1978, *Homosexuals: A Study of Diversity Among Men and Women*, Simon and Schuster, New York.
7. Bell, A., Weinberg, M., and Hammersmith, S. K.: 1981, *Sexual Preference: Its Development in Men and Women*, Indiana University Press, Bloomington.
8. Bieber, I., Davis, H., Dince, P., Drellich, M., Grand, H., Grundlach, R., Kremer,

M., Rufkin, A., Wilber, C., and Bieber, T.: 1962, *Homosexuality: A Psychoanalytic Study*, Basic Books, New York.
9. Bullough, V.: 1976, *Sexual Variance in Society and History*, University of Chicago Press, Chicago.
10. Califia, P.: 1980, *Sapphistry: The Book of Lesbian Sexuality*, Naiad Press, Tallahassee.
11. Chaddock, C. G.: 1894, 'Sexual Crimes', in A. McL. Hamilton and L. Godkin (eds.), *A System of Legal Medicine*, Vol. 2, E. B. Treat, New York, pp. 525–72. Abstracted in Katz [32], pp. 256–257.
12. Cochran, W. G., Mosteller, F., and Tukey, J. W.: 1953, 'Statistical Problems of the Kinsey Report', *Journal of the American Statistical Association* **48**, 673–716.
13. Childs, A.: 1977, 'Acute Symbiotic Psychosis in a Postoperative Transsexual', *Archives of Sexual Behavior*, 6/1, 37–44.
14. Dana, C. L.: 1891, 'Clinical Lecture On Certain Sexual Neuroses', *Medical and Surgical Reporter* (Philadelphia), 65/7, 241–245. Abstracted in Katz [32], p. 223.
15. D'Emilio, J.: 1983, *Sexual Politics, Sexual Communities: The Making of a Homosexual Minority in the United States, 1940–1970*, University of Chicago Press, Chicago.
16. Daniel, F. E.: 1912, 'Castrated "Sexual Perverts"', published at least four places between 1893 and 1912 including *Texas Medical Journal*, 27/10, 369–85. Abstracted in Katz [32], pp. 241–243.
17. Duffy, John: 1982, 'The Physician as a Moral Force in American History', in W. B. Bondeson, H. T. Engelhardt, Jr., S. F. Spicker, and J. M. White (eds.), *New Knowledge in the Biomedical Sciences*, D. Reidel, Dordrecht, pp. 3–21.
18. Duffy, J.: 1963, 'Masturbation and Clitoridectomy: A Nineteenth-Century View', *Journal of the American Medical Association* **186**, 246–248.
19. Duffy, J.: 1984, 'American Perceptions of the Medical, Legal, and Theological Professions', *Bulletin of the History of Medicine* **58**, 1–15.
20. Duffy, J.: 1987, 'Sex Society, Medicine: An Historical Comment', in Earl E. Shelp, (ed.), *Sexuality and Medicine: Ethical Viewpoints in Transition*, D. Reidel, Dordrecht, in press.
21. Engelhardt, H. T., Jr.: 1974, 'The Disease of Masturbation: Values and Concepts of Disease', *Bulletin of the History of Medicine* **47**, 391–409.
22. Ellis, H.: 1895, 'Sexual Inversion: With an Analysis of Thirty-Three New Cases', *Medico-Legal Journal* (New York) **13**, 255–67. Abstracted in Katz [32], pp. 287–288.
23. Fellman, A. C. and Fellman, M.: 1981, 'The Rule of Moderation in Late Nineteenth Century American Sexual Ideology', *Journal of Sex Research*, **17**/3, 238–255.
24. Foucault, M.: 1980, *The History of Human Sexuality*, Vol. I., translated by Robert Hurley, Vintage Books, New York.
25. Frank, J. D.: 1959, 'Problems of Controls in Psychotherapy as Exemplified by the Psychotherapy Research Project of the Phipps Psychiatric Clinic', in E. A. Rubinstein and M. B. Parlott (eds.), *Research in Psychotherapy*, American Psychological Association, Washington, D.C., pp. 10–26.
26. Greene, G. and Greene, C.: 1974, *S-M: The Last Taboo*, Grove Press, New York.
27. Haller, J. S. and Haller, R. M.: 1974, *The Physician and Sexuality in Victorian America*, University of Illinois Press, Urbana.

28. Hamilton, A. McL.: 1894, 'Insanity in its Medico-Legal Bearings', in A. Hamilton and L. Godkin (eds.), *A System of Legal Medicine*, Vol. 2, E. B. Treat, New York, pp. 49–50. Abstracted in Katz [32], p. 25.
29. Henry, G. W. and Galbraith, H. M.: 1934, 'Constitutional Factors in Homosexuality', *American Journal of Psychiatry*, **90**/2, 1249–1270.
30. Jorgensen, Christine: 1953, 'The Story of My Life', in five parts. *The American Weekly*, Part I, Feb. 15, pp. 4–5, 7–9; Part II, Feb. 22, pp. 4–6, 12; Part III, March 1, pp. 14–18; Part IV, March 8, pp. 3–4, 11, 21; Part V, March 15, pp. 10, 13, 15.
31. Katz, J. N.: 1976, *Gay American History: Lesbians and Gay Men in the U.S., A Documentary*, Thomas Y. Crowell Co., New York.
32. Katz, J. N.: 1983, *Gay/Lesbian Almanac: A New Documentary*, Harper & Row, New York.
33. Kiernan, J. G.: 1892, 'Responsibility in Sexual Perversion', *Chicago Medical Recorder* **3**, 185–210; abstracted in Katz [32], pp. 231–232.
34. Kinsey, A., Pomeroy, W., and Martin, C.: 1948, *Sexual Behavior in the Human Male*, W. B. Saunders, Philadelphia.
35. Kinsey, A., Pomeroy, W., Martin, C., and Gebhard, P.: 1953, *Sexual Behavior in the Human Female*, W. B. Saunders, Philadephia.
36. Lester, D.: 1975, *Unusual Sexual Behavior: The Standard Deviations*, Charles C. Thomas, Springfield, Illinois.
37. Livingood, J. M.: 1972, *National Institute of Mental Health Task Force on Homosexuality: Final Report and Background Papers*, DHEW Publication No. (HSM) 72–9116, U.S. Government Printing Office, Washington, D.C.
38. Lothstein, L. M.: 1980, 'The Postsurgical Transsexual: Empirical and Theoretical Considerations', *Archives of Sexual Behavior*, **9**/6, 547–564.
39. Ovesey, L.: 1969, *Homosexuality and Pseudohomosexuality*, Science House, New York.
40. MacKenzie, K. R.: 1978, 'Gender Dysphoria Syndrome: Towards Standardized Diagnostic Criteria', *Archives of Sexual Behavior*, **7**/4, 251–262.
41. Morgan, A. J., Jr.: 1978, 'Psychotherapy for Transsexual Candidates Screened Out of Surgery', *Archives of Sexual Behavior*, **7**/4, 273–84.
42. O'Carroll, T.: 1982, *Paedophilia: The Radical Case*, Alyson, Boston.
43. Paul, W., Weinrich, J. D., Gonsiorek, J. C., and Hotvedt, M. (eds.): 1982, *Homosexuality: Social, Psychological, and Biological Issues*. Sage Publications, Beverly Hills.
44. Prince, V.: 1978, 'Transsexuals and Pseudotranssexuals', *Archives of Sexual Behavior* **7**, 263–272.
45. Person, E. and Ovesey, L.: 1974, 'The Transsexual Syndrome in Males: I. Primary Transsexualism', *American Journal of Psychotherapy* **28**, 4–20.
46. Sandfort, T.: 1982, *The Sexual Aspect of Paedophile Relations*, Pan/Spartacus, Amsterdam.
47. Shively, M. and DeCecco, J.: 1977, 'Components of Sexual Identity', *Journal of Homosexuality* **3**, 41–8.
48. Spengler, A.: 1979, *Sadomasochisten und ihre Subkulturen*, Campus Verlag, Frankfurt/Main.
49. Suppe, F.: 1981, 'The Bell and Weinberg Study: Future Priorities for Research on Homosexuality', *Journal of Homosexuality* **6**, 69–97. Reprinted in N. Koertge (ed.),

Nature and Causes of Homosexuality: A Philosophic and Scientific Inquiry, Haworth Press, New York.

50. Suppe, F.: 1982, Review article on R. Bayer, *Homosexuality and American Psychiatry*, *Journal of Medicine and Philosophy* **7**, 375–381.
51. Suppe, F.: 1983, Review of Paul *et al.*, *Homosexuality: Social, Psychological, and Biological Issues*, *The Advocate*, Issue #360, 74–75.
52. Suppe, F.: 1984, 'Curing Homosexuality', in R. Baker and F. Elliston (eds.), *Philosophy and Sex*, 2nd ed., Prometheus Books, Buffalo, pp. 391–420.
53. Suppe, F.: 1984, 'Classifying Sexual Disorders: The Diagnostic and Statistical Manual of the American Psychiatric Association', *Journal of Homosexuality*, **9**/4, 9–28.
54. Suppe, F.: 1987, 'The Diagnostic and Statistical Manual of the American Psychiatric Association: Classifying Sexual Disorders', in Earl E. Shelp (ed.), *Sexuality and Medicine: Ethical Viewpoints in Transition*, D. Reidel, Dordrecht.
55. Suppe, F.: 1984, 'In Defense of a Multidimensional Approach to Sexual Identity', *Journal of Homosexuality*, **10**/3/4, 7–14.
56. Suppe, F.: forthcoming, 'Ego-dystonic Homosexuality and Sexual Disorders: The Codification of Sexual Mores in the Diagnostic and Statistical Manual of the American Psychiatric Association', in the Proceedings of the 6th World Congress for Sexology.
57. Townsend, L.: 1972. *The Leatherman's Handbook*, Le Salon, San Francisco.
58. Tsang, D., (ed.): 1981, *The Age Taboo: Gay Male Sexuality, Power and Consent*, Alyson, Boston.
59. Weir, James, Jr.: 1895, 'The Effect of Female Suffrage on Posterity', *American Naturalist*, **24**/345, 815–25; abstracted in Katz [32], pp. 285–287.
60. Wilson, P.: 1981, *The Man They Called Monster: Sexual Experiences Between Men and Boys*, Cassell Australia, New South Wales.
61. Wojdowski, P. and Tebor, I. B.: 1976, 'Social and Emotional Tensions During Transsexual Passing', *Journal of Sex Research*, 12/3, 193–205.
62. Wood, G. P. and Ruddock, E. H.: 1909, *Vitalogy, or Encyclopedia of Health and Home; Adapted for Home and Family Use; Beacon Lights for Old and Young, Showing How to Secure Health, Long Life, Success and Happiness, from the Ablest Authorities in this Country, Europe and Japan*, Vitalogy Association. Printed by M. A. Donohue and Company, Chicago.
63. World Health Organization: 1948, *International Classification of Diseases: Manual of the International Classification of Diseases, Injuries, and Causes of Death. Sixth Revision of the International Lists of Diseases and Causes of Death*, Geneva.
64. World Health Organization: 1959, *International Classification of Diseases, Adapted for Indexing of Hospital Records and Operational Classification*, U.S. Government Printing Office, Washington, D.C.
65. World Health Organization: 1962, *International Classification of Diseases, Adapted for Indexing Hospital Records by Diseases and Operations*, revised edition, U.S. Government Printing Office, Washington, D.C.
66. World Health Organization: 1967, *International Classification of Diseases: Manual of the International Statistical Classification of Diseases, Injuries, and Causes of Death*, Eighth Revision, Geneva.
67. World Health Organization: 1977, *International Classification of Diseases: Manual of the International Statistical Classification of Diseases, Injuries, and Causes of Death*, Ninth Revision, Geneva.

STEPHEN B. LEVINE

THE ORIGINS OF SEXUAL IDENTITY: A CLINICIAN'S VIEW

It is not possible to be completely objective about a subject as value-laden and emotionally evocative as sexual preference. Instead of indulging in illusions about my dispassionate, critical thinking in this area, I want to make my biases explicit. The perspective of this essay is clinical. It has evolved over a decade of helping men and women with concerns about their sexual preferences. My clinical bias is developmental. It is based on the assumption that all mental and behavioral phenomena involving sexual preferences are products of the child's integration of biologic, intrapsychic, and social forces. My perspective is also psychodynamic, because I sense that such traditions place more emphasis on the child's unique integration of inherent capacities, internal processes, and social influences than do strictly sociologic, behavioral, or biologic perspectives.

I am mindful of the fact that all clinical traditions share one fundamental scientific limitation – i.e., psychotherapies are retrospective methods. At best, a retrospective method is only capable of generating hypotheses [34]. Psychotherapy involves inferences about causal links between events and processes that occurred in the distant past. The transformation of clinically generated hypotheses into scientific explanations is only accomplished by subsequent retrospective and prospective studies. The generation of similar hypotheses by different clinicians does not constitute proof of their validity.

Finally, my views are influenced by my sexual preferences. I am a masculine, heteroerotic heterosexual with peaceable intentions for mutuality. These characteristics lead to my assumption that other sexual preferences result from less than adequate developmental opportunities. I do not, however, believe that I am determined to prove my assumption correct. I think I am more curious than judgmental about the origins of sexual preference.

E. E. Shelp (ed.), Sexuality and Medicine, Vol. I, 39–54.

THE THREE DIMENSIONS OF ADULT SEXUAL IDENTITY

Terms such as 'sex role,' 'sex role preference,' 'sexual preference,' 'orientation,' 'sexual object,' and 'sexual aim' refer to aspects of sexual identity. The sexual identity of any adult may be described along three separate dimensions: gender identity; sexual orientation; sexual intention. These three dimensions, separately or together, are basically parts of a subjective, psychological, intrapsychic phenomenon. As one facet of a many-faceted sense of the self, sexual identity exists with other identities, such as political, ethnic, religious, generational, vocational. Although there are important objective, behavioral aspects of sexual identity, its uniqueness can be more clearly perceived from its subjective aspects.

Gender Identity

Gender identity is the first aspect of sexual identity to form. The child develops a sense of being a boy or girl early in the second year of life – probably based upon an inconspicuous, repetitive, labeling process underway since birth. The child is taught its gender and subtly steered in various directions by the family (the society of early life). Families have many conventional, and some very unique, attitudes about appropriate behaviors for boys and girls. By accepting their labels and the early steering, most children 'choose' to be further influenced in a masculine or feminine direction. Normal children, between the ages of one and one-half and three, show many subtle signs of temporary gender confusion and envy of the opposite sex [16, 24, 37]. Perhaps as many as ninety percent of children develop what is known as a core gender identity, consonant with their biologic sex, by the middle or end of the third year of life [30]. In recalling early childhood, most parents and children seem to completely forget the instances of cross-gender identifications, curiosities, envy, and fantasy that are often part of the 'cuteness' of early life. This forgetting or inattention to cross-gender preoccupations misled many people into thinking that gender sense is solely a biological phenomenon that simply unfolds, rather than a feeling that the child acquires and elaborates on. The establishment of core gender identity is based on identifications with others, particularly those with whom the child has a trusting relationship. As a consequence of establishing core gender identity at the usual time, boys and girls will maintain a relatively consistent sense of themselves as boys or girls and become preoccupied with behaving in an acceptable masculine or feminine fashion. Because of this vital consequence, core gender identity can be considered the psychological foundation of sexuality.

Childhood forms of femininity must not be confused with the larger sociologic agonies associated with culture's attempt to redefine appropriate sex roles. Once comfortably and inconspicuously established, gender identity continues to evolve and be influenced by the person's subsequent identifications in an ever widening world. At any stage of life, the unique sense of the self as a certain type of male or female can be considered a reflection of gender identity. It becomes progressively harder, however, to define the limits of normal gender identity as humans progress from the rudimentary boy vs. girl identification stage to adult styles of masculinity and femininity. Cultural forces – e.g., television, playmates, educational systems, adolescent subcultures, impact on the child to shape the evolution of gender self images. It is especially difficult to be certain of gender identity development during adolescence. Beyond adolescence, 'normal' generally refers to the absence of concern about masculinity or femininity.

The gender identity of others can be inferred from their behaviors. The rough and tumble play of a four-year-old boy, for example, often reassures parents that their son is 'all boy.' The preference for playing with dolls similarly reassures some parents that their daughter is feminine. Play, clothing style, mannerisms, and interest pattern may be considered gender role behaviors to the extent that they are indicative of a person's gender identity. However, gender role behavior can never be completely relied on to indicate the subjective, psychologic sense of gender identity. Conventional gender role behaviors are often seen in adults with unconventional gender identities. Some adolescents try out many gender roles in the process of consolidating a stable gender identity.

The classic adult disorders of gender identity – i.e., transsexualism; transvestitism; atypical gender disorder – provoke great consternation among lay and professional persons [17]. Many are initially certain that people are crazy (psychotic) when they insist that they are trapped in the wrong body. Biologic sex and psychologic gender are so fundamentally related for conventional persons that even agnostics tend to experience their gender identities as God-given, fundamental aspects of the natural order of the universe. The simple boy-girl beginnings of gender identity develop into far more complicated outcomes. There are well-integrated masculine and feminine identities ('I know I am like a woman in some ways, but I nonetheless feel comfortably masculine'). There are compartmentalized masculine and feminine identities ('I feel like a man sometimes, but live as a female when I can'). There are also neuter or ambiguous identities ('I don't know what I am!'). Healthy adult gender identity is rich with nuance, subtlety, and integration. Adult gender identity disorders represent the failures to comfortably resolve the problem every toddler faces: 'Am I a boy or girl? Is that all right with me?'

Sexual Orientation

The second dimension of sexual identity is sexual orientation, which also has subjective and objective aspects. Adult subjective orientation refers to the sex of people or mental images of people that attract and provoke sexual arousal. Adults can be considered heteroerotic if the vast majority of images, fantasies, and attractions associated with sexual arousal concern members of the opposite sex. Homoerotic adults think about, are attracted to, or are aroused by images of persons of the same sex. Bierotic individuals have the ability to become sexually aroused by images of both sexes. Sexual orientation is also reflected in adult behavior. Adults can be classified as heterosexual, homosexual, or bisexual – depending on the biologic sex of their partners. The two aspects of sexual orientation are not always consistent: a homoerotic woman may only behave heterosexually; a man who behaves bisexually may be entirely homoerotic; a heteroerotic man may engage in homosexual behavior.

The homoerotic nature of male orientation is usually consciously manifested several years before heteroerotic orientation. Onset of partner sexual behavior and masturbation also tends to be earlier in homoerotic grade school and junior high students. The opposite patterns tend to be true for homoerotic and heteroerotic girls – i.e., female homoerotic orientation tends to manifest later than female heteroeroticism [3, 25].

Sexual Intention

Sexual intention — what a person actually wants to do with his or her sexual partner – constitutes the final dimension of sexual identity.

Conventional sexual intentions include a wide range of behaviors, such as kissing, caressing, genital union, which are mutually pleasurable to consenting persons. Conventional intentions involve giving and receiving pleasure. While the behavioral repertoire of conventional intention is usually wide, the fantasy repertoire may be much wider. Unconventional sexual intentions involve raw or disguised aggression toward a victim, rather than mutual pleasure. The victim may be the self or another person. In addition, they are often relatively limited to a few behaviors that provoke arousal. Sadism, masochistic degradation, exhibitionism, voyeurism, rape, and pedophilia are examples of unconventional developmental outcomes of intention. Erotic intention refers to the intrapsychic fantasy aspects. Behavioral intention refers to what is actually acted out. Unconventional erotic intentions, generally referred to as paraphilias, are far more common than unconventional behavioral intentions. Unconventional intentions are probably first manifested with clarity for most individuals in adolescence. However, careful review of individual fantasy evolution often demonstrates the presence of precursors for many years [20, 32].

Each of the three dimensions of sexual identity can be thought of as structures of the mind created through the processes of development. The structures remain relatively fixed throughout adulthood. In general, gender identity is more fixed than sexual orientation, which, in turn, is more enduring than sexual intention. The constancy of the structures is not absolute, however. Dramatic shifts in each dimension of sexual identity have been reported [1, 8, 18], and many unreported shifts have been observed clinically. Nonetheless, the relative stability of these structures should be emphasized. Table I summarizes the identity

TABLE I
Sexual identity structures of the adult conventional mind

	Male	*Female*
gender identity	masculine	feminine
gender role	masculine	feminine
erotic orientation	heteroerotic	heteroerotic
sexual orientation	heterosexual	heterosexual
erotic intention	peaceable mutuality	peaceable mutuality
behavioral intention	peaceable mutuality	peaceable mutuality

structures of the conventional mind. Sexual identity structures are clinically manifested in reports of attractions, fantasies, partner behavior, and dreams. The clarity of the conventional person's classification is often subtly obscured during intensive psychotherapy. The intimacy of the therapeutic process often reveals unconventional elements imbedded in conventional structures. A clinician who has this experience is better able to accept the unverifiable idea that few, if any, people are entirely conventional in their attractions, fantasies, dreams, or behavior over a lifetime. It is, then, less shocking to realize that a 'normal' feminine, heteroerotic, heterosexual woman with peaceable intentions may, for example, occasionally experience homoerotic images, have fantasies of victimizing her partner, or have a sense of being masculine. Clinicians should not be taken aback to discover that a masculine heterosexual rapist evidences feminine identification, homoeroticism, and masochistic preoccupation in his subjective life. The subjective side of sexual identity is quite private and generally denied to scientists who probe with questionnaires and rating scales, as well as to clinicians most of the time.

Two Caveats

The stability of sexual identity structures is a recurrent, vital, clinical question – i.e., can this transsexual (transsexualism), homosexual (homosexuality), masochist (masochism), rapist (perversion) be cured? Such questions need to be considered in the larger context of the stability of other personality structures. Once they have been programmed, are these structures imbedded in the unconscious mind, inaccessible to environmental influences, and changeable only in response to the mysterious process of maturation? Is the stability of sexual identity due to the fact that people select relatively consistent environments over time [11]? Sexual identity structures can and do adapt in response to life events [7, 23, 29], just as nonsexual aspects of personality sometimes evidence dramatic changes in response to unexpected social circumstances. Asking whether cross-dressers or exhibitionists can be cured is not that different from asking whether obsessive-compulsive, phobic avoidance patterns can be cured. One must be careful to avoid the trap of considering sexual identity a special case.

However useful they may be in indicating general trends, words such as 'masculine,' 'heterosexual,' and 'masochistic' must not be considered

TABLE II
Some unconventional sexual identity structures

Biologic Sex	*Male*	*Female*	*Male*	*Female*
gender identity	masculine	ambiguous	feminine	feminine
gender role	masculine	feminine	masculine	feminine
erotic orientation	homoerotic	homoerotic	homoerotic	bierotic
sexual orientation	bisexual	homosexual	heterosexual	bisexual
erotic intention	masochistic	sadistic	silk fetish	exhibitionistic
bahavioral intention	personal degradation	mutuality	self-centered cross-dressing	masochistic

too specific in their delineations [33]. Sexual identity structures are evolving dynamic phenomena. The possible combinations of gender identity, sexual orientation, and intention are quite numerous; in fact, there are probably more than were previously realized. Table II illustrates just a few of the many unconventional forms that exist at any given point in time. Unconventional structures may shift considerably in a short period of time.

INCOMPLETELY ANSWERED QUESTIONS ABOUT SEXUAL IDENTITY

1. What are the essential biological, social, and psychological processes required to produce conventional adult sexual identity structures?
2. When are the templates for gender identity, orientation, and intention individually formed in the child's mind?
3. What are the likely adult consequences of delayed formation of these templates?
4. What is the relative importance of the timely establishment of these templates to the influence of adolescent and early adult experience?
5. Do specific forms of unconventional sexual identity structures – i.e., masochism, sadism – stem from specific sources?
6. Should unconventional sexual identity structures be considered as isolated problems, or do they reflect more basic developmental difficulties?

HOMOSEXUALITY: THE MOST THOROUGHLY STUDIED UNCONVENTIONAL IDENTITY STRUCTURE

In 1975, a thorough review of the research literature on the sources of homosexual behavior suggested the field "return to the drawing board" [2]. Subsequently, a number of lines of evidence have been converging to suggest a new hypothesis: perhaps as many as two-thirds of adult male homosexuals had active, conscious wishes to be females for prolonged periods of childhood. In other words, homoerotic orientation and homosexual behavior are adolescent and adult outcomes of earlier cross gender identifications and aspirations. A number of retrospective studies of adult homosexual men have revealed memories of 'girl like' behavior and aspirations in early grade school [12, 35, 38]. After carefully refining the retrospective method, Whitam confirmed this hypothesis [39] with evidence from homosexual males in three cultures [40]. The identified characteristics of the prehomosexual child were:

1. use of toys of the opposite sex;
2. cross-dressing;
3. preference for girls' games, activities, and company of women;
4. being regarded as a sissy;
5. preference for boys in childhood sex play.

Meanwhile, prospective follow-up studies of feminine preschool boys indicated that the usual adolescent outcome was homoerotic homosexual orientation. Rarer outcomes were transvestitism, transsexualism, and heterosexuality [13, 41]. Recent research has suggested that adult male homosexuals make a dramatic, late latency-early adolescent developmental effort to defeminize themselves [14]. The deliberate process of eliminating effeminate traits and mannerisms makes it possible for the majority of femininized prehomoerotic males to end up with a grossly masculine gender identity and role; their homoerotic orientation remains.

This hypothesis enables the question of the causes of homosexuality to be rephrased into, 'What are the biologic, social, and psychologic sources of prolonged, predominant cross gender identifications?' The research into transsexual and homoerotic phenomena now seem to be more relevant to each other (see Lothstein's essay in this volume).

In the past three decades, a great deal of effort has been devoted to demonstrating a biologic contribution to transsexual and homosexual phenomena. It is very hard to reject the notion that the heterosexual behavior that ensures species perpetuation is not biologically dictated by genetic, neuroendocrine influences. Yet, the weight of the evidence suggests that in humans the chromosomal, foetal gonadal, foetal hormonal, and foetal hypothalamic differentation influences can easily be overridden by the postnatal social influences initiated by gender assignment [21]. The parents' perception of the child's sex – 'It's a boy!' 'It's a girl!' — ushers in an increasingly understood social-psychologic process that results in the child's acquisition of a gender identity [23]. Gender identity is either appropriate or inappropriate to the genital structure. As development proceeds, the cultural influences on core gender identity play an ever-increasing, yet always subtle, role. There is little replicated evidence to suggest that biological processes have a major role in directing gender identity. Occasionally, however, case histories of poor adjustment in gender-assigned roles again suggest the intuitively reasonable notion that nature is a more powerful determinant than nurture (at least in some individuals!) [10, 30].

The psychoanalytic hypotheses about male homosexuality have often been dismissed because of methodologic limitations inherent in the data generation. Bieber *et al.* interpreted the child's fear of rough and tumble aggressive play as evidence of too close a relationship to the mother and paternal indifference [5]. Socarides refined concepts of the origin of male homosexuality into two types: preoedipal, in which prolonged closeness-delayed separation-individuation phases led to intense feminine identifications; oedipal, in which early masculine identifications became complicated in response to the sexual triangle in which the three-to-five-year-old male found himself enmeshed [28]. While such conceptual clarity may be useful, the preoedipal contributions to the style of oedipal phase conflict resolution make it unrealistic. The earlier formulations of Bieber *et al.* and Socarides' continuing refinements of hypotheses are in agreement with the emerging central hypothesis that two-thirds of male homosexuals obtained their homoerotic orientation as a result of early life cross gender identifications.

The strength of the central hypothesis about the precursors of male homoeroticism raises the comparable question about female homoeroticism. Are the persistent homoerotic phenomena of late adolescence and young adulthood the consequence of earlier prolonged identifications with, and aspirations to be, males? Several studies have confirmed the much higher prevalence of tomboyism among adult lesbians than heterosexuals [3, 15, 25]. Female homosexual phenomena should not, however, be considered the mirror image of male homoeroticism. Along most parameters, female homosexuals are more like female heterosexuals than male homosexuals [3, 25]. Considerable uncertainty exists about female homosexuality because its origins have rarely been studied. Moreover, female homosexuals seem to avoid traditional mental health professionals. Even clinical data are conspicuously sparse.

There are many homoerotic males and females whose histories do not suggest prolonged periods of early cross gender identification. These situations are usually explained by reminders about the multiple developmental pathways to homosexual behavior or the fact that man's basic nature is bisexual. Such aphorisms may not be as helpful as more specific questions: What conditions enable the appearance of homoeroticism after core gender identity has been conventionally established? Are there particular family dynamics, personality traits, or social situations that facilitate the appearance of homoeroticism? (Homoeroticism, not homosexuality! Social isolation and power politics can cause heteroerotics to behave homosexually.) Does the persistent belief that he can't compete against his loved father, is hated by him, or is deprived of needed attention and association predispose a five-year-old to adolescent homoeroticism? Can the emerging awareness of a mother's limited capacity to love cause a feminine daughter to develop a hunger for a new mother that will emerge years later as a homoerotic sexual identity? Can a very positive, loving, nonsexual relationship with a homosexual adult lead, via identification, to homoeroticism? Heterosexual women who are profoundly committed to feminism sometimes engage in homosexual behavior because it represents an ideologic purity [9]. Such observations raise the question of the impact of actual homosexual experience on emerging orientation structures.

Given our uncertainty about the determinants of orientation structures, it is important to realize that each phase of life may influence the final outcome. Traditionally, clinicians have looked for explanations in early childhood periods; they assumed that adolescence was a reworking of early childhood issues at a higher level [6]. The current psychoanalytic theory of sexual development is, however, much more open-minded. Sexual identity is thought to be influenced by contributions from oral, anal, phallic-narcissistic, phallic oedipal, latency, and early, middle, and late adolescent stages of development [36]. The continuity of the formative processes is emphasized. The preoedipal stages in both boys and girls are now weighted most heavily [19, 29]; in the past, the oedipal stage was emphasized.

SEXUAL INTENTION DISORDERS

Clinical attention and research only rarely focus on the origins of specific sexual intentions. It is agreed that the templates of gender identity and orientation are usually set by age six; we do not know when and how a specific form of intimate contact becomes eroticized.

Explanations for the attainment of conventional intentions must at least consider the widespread social norm of cooperative, willing partners who derive emotional satisfaction from sexual behavior. Individuals are socialized toward this goal. Moreover, the individual's erotic intention does not necessarily determine actual sexual behavior with partners. Sexual behavior is reality between people, not simply fantasy. The actual behavior forces most individuals to experience the partner's humanity. Even in the many sexual relationships that fall short of this ideal – e.g., prostitution – one can sometimes detect the man's attempt to preserve the mutuality of the experience through fantasy.

The attainment of conventional subjective and behavioral intentions is a developmental landmark, for it frees the individual to deal with relationships at a higher level of complexity. When conventional people have unconventional fantasies, they are usually in response to a highly upsetting life experience or an unusual visual exposure to erotica. Such experiences typically 'capture' the conventional mind for brief periods because they involve fantasies that offend the person's standards for personal behavior. The following case illustrates the point:

Case 1:
A 30-year-old man became very anxious after masturbation to the image of defecating on his partner's face during lovemaking. This fantasy was created one day after he realized his girlfriend was manipulating him in order to advance her career. Discussing his anger at her callous indifference to his feelings allowed him to end their association; his intentions rapidly returned to their conventional state.

This conventional man's mind was briefly dominated by vengefulness. He needed to provide a humiliation commensurate with his own disgrace. The vengefulness arose out of an intimate relationship.

Fortunately for him, the traumatic episode occurred when he was well into his adulthood.

This ordinary human experience may assist in the understanding of the two more difficult forms of paraphiliac mind capture: paraphiliac fantasies that inhibit partner behavior; paraphiliac fantasies that are acted on with partners.

Case 2:
A 48-year-old woman who could only achieve orgasm by masturbating to an enema fantasy had difficulty enjoying any sexual behavior during twenty-five years of marriage. She experienced her fantasy about losing control of the contents of fecal matter as sick. After years of therapy she recalled that her parents had forced her to have daily enemas for a week when she was seven. Her mother suddenly died one month later. As the details of the fantasy became clearer, she realized that the fecal fluid was always spattered on her parents. The fantasy disguised her need to avenge the recurrent violation of her inner space. By capturing her mind, this fantasy allowed her to triumph over her humiliating loss of self control: it also kept her preoccupied with her mother, as if she were a living person. The fantasy expressed her rage over being violated, guilt over her anger, and confusion about the cause of her mother's death.

Case 3:
A 25-year-old homoerotic man had employed two masturbatory fantasies since age 9. In the first, he was tied up, undressed and anally penetrated against his will. This fantasy alternated with one in which he tied up a young man, undressed him, and watched him being gang raped. His orgasm was usually triggered by his or the victim's shouting, 'No don't!' This theme remained enigmatic during several years of therapy until he remembered an experience from his fifth year of life. In the hospital awaiting surgery for a congenital orthopoedic deformity, he ignored the first request to go to the operating room. When the nurse returned, she slyly asked him if he'd like to go for a ride on a cart. Tricked into getting on the cart, he recalled being strapped down, undressed by strangers, and terrified at having the ether mask placed over his face. All the while he begged 'No don't, please don't!'

Clinical experience suggests that minds can be captured by paraphiliac images at any age. The essence of the experience may be described as powerlessness, dependency, and the sense of being abused. Stoller's hypothesis seems correct – i.e., perversion represents the child's rearrangement of the traumatic situation so that he or she triumphs [31]. The child may eroticize the victim experience in order to master it. This makes perverse eroticization a special form of defense, usually described as turning passive-into-active.

Unfortunately, the traumatic triggers of many perverse fantasies and behaviors cannot be ascertained. Events, *per se*, may not be as traumatic as processes; the general poor fit between the needy child and the emotionally unavailable parent is traumatic. Moreover, those who carry out their unconventional acts with victims, rather than consenting partners, are rarely called patients; they are called sex offenders or criminals. The few cases that have been treated suggest that there is a direct relationship between bizarreness, persistence and intensity of the perverse image, and the global nature of the personality problems. Thus, men who masturbate 10 times a day to the image of committing rape or being mutilated generally have an array of psychiatric diagnoses other than paraphilia. It is possible that the severity of the paraphilia is determined by subtle defects in neural integration [4].

SUMMARY

This essay has provided a developmental clinical perspective on sexual identity. The study of sexual identity is actually an emerging frontier of sex research and psychoanalysis. Both the vocabulary of the field and its etiologic concepts are changing – current theories about healthy and pathological development are extremely complex [26]. The intrapsychic process of the individual mind is no longer the unit of study; instead, the focus is on the social field, which is defined differently at different stages of life – e.g., mother-infant; parents-child; family-child; family-culture-child; culture-child; family-culture-adolescent; adolescent-adolescent; adult-adult; etc. Awareness of the basic complexity of the developmental processes has led to the realization that there are no simple answers.

This essay has not dealt with the relationship between the developmental sexual identity structures and other personality capacities and structures. Clearly, sexual identity patterns are not established in isolation from object constancy, separation-individual processes, cognitive development, mood regulation, task mastery, and a myriad of other continuing developmental complexities. In evaluating patients with sexual identity problems, it is important to consider the dimensions of their sexual identities in relation to other aspects of their lives.

Case Western Reserve University,
Cleveland, Ohio,
U.S.A.

ACKNOWLEDGEMENT

The author thanks Ms. Phyllis Polsky and Mrs. Barbara Juknialis of the Department of Psychiatry for their invaluable secretarial and editorial assistance in the preparation of this essay.

BIBLIOGRAPHY

1. Barlow, D. H. *et al.*: 1977, 'Gender Identity Change in a Transsexual: An Exorcism', *Archives of Sexual Behavior* **6**, 387–396.
2. Bell, A.: 1975, 'Research in Homosexuality: Back to the Drawing Board', *Archives of Sexual Behavior* **4**, 421–432.
3. Bell, A. and Weinberg, S.: 1978, *Homosexualities: A Study of Diversity Among Men and Women*, Simon and Schuster, New York.
4. Berlin, F. S. and Meinecke, C. F.: 1981, 'Treatment of Sex Offenders with Antiandrogenic Medication: Conceptualization, Review of Treatment Modalities and Preliminary Findings', *American Journal of Psychiatry* **135**, 601–607.
5. Bieber, I. *et al.*: 1962, *Homosexuality: A Psychoanalytic Study*, Basic Books, New York.
6. Blos, P.: 1979, *Adolescent Passage*, International Press, New York.
7. Bobys, R. S. and Laner, M. R.: 1979, 'On the Stability of Stigmatization: The Case of Ex-homosexual Males', *Archives of Sexual Behavior* **8**, 247–261.
8. Davenport, C. W. and Harrison, S. I.: 1977, 'Gender Identity Change in a Female Adolescent Transsexual', *Archives of Sexual Behavior* **6**, 327–340.
9. Defries, Z.: 1976, 'Pseudohomosexuality in Feminist Students', *American Journal of Psychiatry* **133**, 400–404.
10. Diamond, M.: 1982, 'Sexual Identity, Monozygotic Twins Reared in Discordant Sex Roles and a BBC Follow-up', *Archives of Sexual Behavior* **11**, 181–186.
11. Eisenberg, L.: 1977, 'Development as a Unifying Concept in Psychiatry', *British Journal of Psychiatry* **131**, 225–237.
12. Freund, K. *et al.*: 1974, 'Measuring Feminine Gender Identity in Homosexual Males', *Archives of Sexual Behavior* **3**, 249–261.
13. Green, R.: 1979, 'Childhood Cross Gender Behavior and Subsequent Sexual Preference', *American Journal of Psychiatry* **135**, 692–697.
14. Harry, J.: 1983, 'Defeminization and Adult Psychological Well-being Among Male Homosexuals', *Archives of Sexual Behavior* **12**, 1–20.

15. Kaye, H. E. *et al.*: 1967, 'Homosexuality in Women', *Archives of General Psychiatry* **17**, 626–634.
16. Kestenberg, J.: 1956, 'Vicissitudes of Female Sexuality', *Journal of The American Psychoanalytic Association* **4**, 453–476.
17. Lothstein, L. M.: 1982, 'Sex Reassignment Surgery: Historical, Bioethical and Theoretical Issues', *American Journal of Psychiatry* **139**, 417–426.
18. Lothstein, L. and Levine, S.: 1981, 'Expressive Psychotherapy with Gender Dysphoria Patients', *Archives of General Psychiatry* **38**, 924–929.
19. Mahler, M. *et al.*: 1975, *The Psychologic Birth of the Human Infant*, Basic Books, New York.
20. Meyer, J. and Dupkin, C.: 1983, 'Sadomasochism', in W. E. Fann *et al.* (eds.), *Phenomenology and Treatment of Psychosexual Disorders*, Spectrum, New York, pp. 13–21.
21. Meyer-Bahlburg, H.: 1977, 'Sex Hormones and Male Homosexuality in Comparative Perspective', *Archives of Sexual Behavior* **6**, 297–325.
22. Money, J. and Ehrhardt, A.: 1972, *Man and Woman, Boy and Girl*, Johns Hopkins Press, Baltimore.
23. Pattison, E. M. and Pattison, M. L.: 1980, '"Ex-Gays": Religiously Mediated Change in Homosexuals', *American Journal of Psychiatry* **137**, 1553–1562.
24. Ross, J. M.: 1975, 'The Development of Paternal Identity: A Critical Review of the Literature on Nurturance and Generativity in Boys and Men', *Journal of the American Psychoanalytic Association* **23**, 783–818.
25. Saghir, M. and Robins, E.: 1973, *Male and Female Homosexuality: A Comprehensive Investigation*, Williams and Wilkins, Baltimore.
26. Scharff, D. E.: 1982, *The Sexual Relationship: An Objective Theory View of Sex and the Family*, Routledge and Kegan Paul, Boston.
27. Schwartz, M. F. and Masters, W. H.: 1984, 'The Masters and Johnson Treatment Program for Dissatisfied Homosexual Men', *American Journal of Psychiatry* **14**, 173–181.
28. Socarides, C. W.: 1974, 'Homosexuality', in S. Arieti and E. B. Brody (eds.), *American Handbook of Psychiatry III*, Basic Books, New York, pp. 291–315.
29. Socarides, C. W.: 1979, 'A Unitary Theory of the Perversions', in B. Karasu and C. W. Socarides (eds.), *On Sexuality*, International Universities Press, New York, pp. 161–188.
30. Stoller, R. J.: 1968, *Sex and Gender*, vol. 1, *The Development of Masculinity and Femininity*, Jason Aronson, New York.
31. Stoller, R. J.: 1975, *Perversion, The Erotic Form of Hatred*, Pantheon Books, New York.
32. Stoller, R. J.: 1979, *Sexual Excitement: The Dynamics of Erotic Life*, Pantheon, New York.
33. Stoller, R. J.: 1980, 'Problems with the Term "Homosexuality"', *Hillside Journal of Clinical Psychiatry* **2**, 3–25.
34. Thomas, A. and Chess, S.: 1984, 'Genesis and Evolution of Behavioral Disorders: From Infancy to Early Adult Life', *American Journal of Psychiatry* **141**, 1–9.
35. Thompson, S. K. and Bentler, P. M.: 1973, 'A Developmental Study of Gender Constancy and Parent Preference', *Archives of Sexual Behavior* **2**, 379–385.

36. Tyson, P.: 1982, 'The Developmental Line of Gender Identity, Gender Role and Choice of Love Object', *Journal of American Psychiatric Association* **30**, 61–87.
37. Van Leeuwen, K.: 1966, 'Pregnancy Envy in the Male', *International Journal of Psychoanalysis* **47**, 319–324.
38. Werner, D.: 1979, 'A Cross-cultural Perspective on Theory and Research on Male Homosexuality', *Journal of Homosexuality* **4**, 345–362.
39. Whitam, F. L.: 1977, 'Childhood Indicators of Male Homosexuality', *Archives of Sexual Behavior* **6**, 89–96.
40. Whitam, F. L.: 1980, 'The Pre-homosexual Male Child in Three Societies: The United States, Guatemala, Brazil', *Archives of Sexual Behavior* **9**, 87–99.
41. Zuger, B.: 1978, 'Effeminate Behavior Present in Boys from Childhood: Ten Additional Years of Follow-up', *Comprehensive Psychiatry* **19**, 363–369.

LESLIE M. LOTHSTEIN

THEORIES OF TRANSSEXUALISM

INTRODUCTION

Although transsexualism is recognized as 'un mal ancien', which has historical, mythological, cultural and anthropological roots, it was not until 1980 that it was formally recognized by the American Psychiatric Association as a serious emotional disorder [1]. In fact, the term, 'transsexual,' only appeared in the literature in 1949 [9] and it was not until 1966 that it was accorded clinical status by Benjamin [4] whose pioneering work, *The Transsexual Phenomenon*, provided the first textbook on transsexualism. In this sense, one must regard the clinical disorder of transsexualism as a recent phenomenon.

Over the past two decades, however, the phenomenon of transsexualism has attracted widespread interest; not only as a clinical entity but also as a social, literary, cinematographic, scientific, poetic, surgical, neurohormonal, and mystical phenomenon. As a result of converging evidence on transsexualism from many disciplines our understanding of the etiology, pathogenesis, and phenomenological and theoretical aspects of male and female transsexualism have undergone considerable revision. Moreover, as a consequence of large numbers of surgically sex changed individuals entering the mainstream of society (between 3000 to 6000 worldwide) serious questions have been raised about the meaning and implication of sex reassignment surgery insofar as it may lead to dramatic changes in such traditional social phenomena as the family, marriage, ritual, religion and legal definitions of sex (while also possibly leading to the evolution of ideas concerning sexual identity and sex roles). Over the past decade a number of theories of transsexualism have been proposed that challenge some of the classical views of the disorder. Because of the relevance of transsexual theory to diagnosis, treatment, and bioethical issues, it would seem that an appraisal of those theories is in order. It is the aim of this essay to outline the newer theories of transsexualism (placing them within the context of the so-called classical theories of transsexualism); a task which may shed

E. E. Shelp (ed.), Sexuality and Medicine, Vol. I, 55–72.

light on some of the critical diagnostic, treatment, and bioethical dilemmas facing the transsexual and his/her society.

PHENOMENOLOGICAL ISSUES OF DIAGNOSIS

During the 19th century physicians, neuropsychiatrists, and sexologists documented case histories of individuals who wished to change their sex [29]. These patients (males and females) were viewed as bisexualists who shared many common features: from early childhood on they wished to become the opposite sex; complained that they were 'a man/woman trapped in a woman/ man's body'; exhibited a compulsive desire to cross dress and live and work in the cross gender role; desired to attract same-sex partners in a love relationship (while denying they were homosexual and viewing the ensuing relationship as a 'heterosexual' one); and were generally viewed as non-psychotic (that is, they were not delusional about their sexual status; they knew that they were a male or female but *wished* to become the opposite sex). Patients who exhibited these core features of a gender identity disturbance were initially labeled according to current standards of psychiatric nomenclature. Over the course of the last 100 years these gender disturbed patients have been variously diagnosed as having contrary sexual feelings; being sexual inverts; having a *Hässlichkeitskomplex*; transvestites; exhibiting metamorphosis *sexualis paranoica*; contrasexists; having *psychopathia transsexualis*; eonism; true transsexuals; having *paranoia transsexualis*; psychosexual hermaphrodites; psychosexual inverts; and as being gender dysphoric. However, all of these patients shared certain commonalities. They had a core belief about being the opposite gender, and an obsessive and relentless drive to change their body to match their mind (through sexual surgery). The core features of transsexualism were eventually incorporated into the third edition of the American Psychiatric Association's Diagnostic and Statistical Manual of Mental Disorders (DSM III). According to DSM III [1] the following criteria must be present for a DSM III diagnosis of transsexualism to be made:

1. sense of discomfort and inappropriateness about one's anatomic sex;
2. wish to be rid of one's own genitals and to live as a member of the other sex;
3. the disturbance has been continuous (not limited to periods of stress) for at least two years;
4. absence of physical intersex or genetic abnormality;

5. not due to another mental disorder such as schizophrenia.

The DSM III criteria for diagnosing transsexualism were based solely on descriptive criteria that were unrelated to the possible etiology of the disorder (a diagnosis that many psychodynamically oriented clinicians felt uncomfortable making since it relied on behaviorally oriented criteria alone). While the DSM III diagnostic enterprise may have aimed at methodological consistency, it ignored the wealth of clinical studies linking transsexualism to unconscious fantasies, psychodynamic precursors, a variety of character problems, and, more recently, to borderline pathology. Moreover, the DSM III diagnosis of transsexualism implied that sex reassignment surgery (SRS) was the treatment of choice (a controversial, if not experimental, procedure) and artifically created a number of serious social, psychological, and bioethical dilemmas.

THEORETICAL ISSUES

Most of the early theories of transsexualism were based on only a few clinical cases; theories that viewed transsexualism as either a functional or an organic disorder. Benjamin [4], summarizing the extant theories of transsexualism through the 1960s, concluded that a combination of organic and functional causes led to transsexualism. He stated:

> Our genetic and endocrine equipment constitutes either an unresponsive, sterile, or a more or less responsive, that is to say, fertile soil on which the wrong conditioning and a psychic trauma can grow and develop into such a basic conflict that subsequently a deviation like transsexualism can result ([4], p. 108).

In this section we shall review the major organic and functional theories of transsexualism.

The view that transsexualism might be the result of an underlying organic condition was appealing for several reasons: it alleviated the guilt of the patients; provided a rationale for surgical intervention; and also provided an exotic hypothesis for behavioral neuroendocrinologists to pursue. But what is the evidence? Is transsexualism caused by a biological disorder?

PSYCHOBIOLOGICAL THEORIES

The research in this area has been quite diverse and at times intriguing. Investigators have focused on the role of genes, chromosomes, enzymes, neurotransmitters, neurohormones, prenatal hormones, and H–Y antigen factors in the etiology of transsexualism. Each era brings a specific biological hypothesis into the public domain as a possible 'cause' of transsexualism. Moreover, there have been a series of reports in which a person diagnosed as a transsexual eventually turned out to have an undiagnosed biological disorder (one patient turned out to have a rare hormone enzyme defect, 17b hydroxy-steroid dehydrogenase deficiency [48]; another transsexual patient was eventually discovered to have a chromosomal abnormality, XO/XX mosaicism). Finally, there is the famous Imperato-McGinley *et al.* [17] study of 38 Dominican male children who were raised as girls but at puberty developed as boys! It was eventually discovered that these 'girls' had a pseudohermaphroditic condition resulting from a genetically determined deficiency of the enzyme Delta^4 steroid 5 alpha-reductase. This deficiency affected their appearance so that as children their genitals appeared female and they were raised as females. At puberty, however, they developed fully as males. Although not everything is known about this group the authors claim that the 'girls' knew all along that they were boys (though they were reared as girls) and had no difficulty identifying as males. Certainly this is an intriguing study, though one which has been questioned by some authorities [43].

In their book *Man and Woman/ Boy and Girl* Money and Ehrhardt [34] attempted to shed light on the effect of all the above variables on typical gender identity development. The book summarizes the research of the effects of all types of biological conditions on the formation of gender role and identity.

But what about the evidence? Is transsexualism a biological disorder? Indeed, there is evidence that some transvestites and transsexuals are motivated because of underlying cerebral pathology, specifically temporal lobe disorders [10]. However, only a small number of transsexuals have actually been diagnosed as having a temporal lobe disorder. In addition, when EEG measures are taken, the results, while occasionally suggestive of a possible organic link with transsexualism, are unimpressive. Blumer [5] concluded that while there might be "an occasional close relationship between sexual aberrations (transvestitism in particular) and paroxysmal temporal lobe disorders . . . that definite EEG abnormalities are not a common occurrence in transsexuals." Späte [45], however, chose to go out on a limb. He stated that EEG abnormalities were the cause of transsexualism in two females that he studied. However, such a conclusion seemed unwarranted by his clinical data. In summary, the EEG findings, while suggestive of a link between cerebral pathology and transsexualism, are inconclusive (especially since EEG abnormalities have been associated with non-organic personality disorders which many transsexuals have [15]).

Another biological hypothesis has focused on the possible chromosomal defects in transsexuals. Again the evidence is inconclusive. The results of karyotyping are generally negative and the expense of the tests has called them into question as a routine part of a transsexual evaluation. Indeed, Hoenig and Torr [16] using the most sophisticated methods of chromosome analysis concluded that in all cases they studied (15 male and 5 female transsexuals) the patients' physiological sex matched their chromosomal sex.

Some transsexuals have been diagnosed as having a serious hormonal disorder: e.g., progestin induced hermaphroditism; adreno-genital syndrome; testicular feminizing syndrome; Turner's and Kleinfelter's syndrome [34]. However, the number of transsexuals who have these disorders are so few, and their symptomology so diverse (often varying significantly between transsexuals), that it seems improbable that transsexualism could be explained by those disorders.

The most promising research area linking transsexualism with hormonal causes is in fetal behavioral neuroendocrinological research, focusing on the effects of prenatal hormones on childhood and adult behavior. Ehrhardt and Meyer-Bahlburg [13], summarizing the findings on the effects of prenatal hormones on gender-related behavior, concluded that: "The evidence accumulated so far suggests that human psychosexual differentiation is influenced by prenatal hormones, albeit to a degree . . . The development of gender identity seems to depend largely on the sex of rearing."

One other area needs to be clarified. Many transsexuals claim to have spontaneous feminization or virilization (as the case may be). It is commonly accepted among gender researchers that whenever a transsexual patient is evaluated for a 'spontaneously' occurring hormonal abnormality, the clinician should suspect that the patient has been secretly administering opposite sex hormones (even if they initially deny doing so).

In conclusion, the evidence from all areas of investigation suggests that while any or all of the suspected biological factors may play a facilitating role in the establishment of transsexualism, there is no hard evidence that transsexualism is caused by organic pathology. However, as new insights are gained into the micromolecular structure and functioning of the endocrine system, we will probably have to modify some of our assumptions about how gender identity and role are formed. The best evidence to date, however, suggests that "the development of gender identity seems to depend largely on the sex of rearing"[13].

PSYCHOLOGICAL THEORIES OF TRANSSEXUALISM

A number of theoretical approaches have been put forth to explain transsexualism as a functional disorder. These approaches embrace a variety of theoretical frameworks, including psychoanalysis [23, 44, 46, 47]; behaviorism and social learning theory [2, 6, 41]; cognitive developmental theory [19, 25]; family dynamics and systems theory [8, 37, 48, 52, 53]; object relations theory [39, 40, 51]; and self psychology [7, 29]. While these theories may seem mutually exclusive, in most instances they are interrelated and are separated in name only. In this section, I will review the major theories of transsexualism. These theories have been derived from clinical studies of adult transsexuals and from observational studies of children who have severe gender identity conflicts and have been unable to consolidate a stable core gender identity.

CLINICAL STUDIES OF TRANSSEXUALISM

Any analysis of transsexualism must start with Stoller's [46, 47] classical theory of transsexualism. Essentially, Stoller postulated two different theories to account for male and female transsexualism respectively. Male transsexualism was seen as a rare disorder affecting only a small subgroup of patients who are self-identified transsexuals; that is, the most feminine appearing males with an early history devoid of any establishment of maleness. Stoller viewed this subgroup as 'true' male transsexualism, a disorder which evolved out of a non-conflictual process in which a boy imprinted a female identity from his mother. Stoller argued that these boys 'never' exhibited a period of masculinity and 'never' valued their penis. As a subgroup of transsexualism these boys were seen as having a similar clinical picture with identical psychodynamics and etiology. The majority of patients who requested sex change were not 'true' transsexuals but individuals with a variety of clinical conditions, who, under stress, regressed and experienced intense anxiety over their gender identity and role.

The picture Stoller presented was as follows. The mothers of the 'true' transsexuals were bisexual; had a flawed femininity; their mothers (the maternal grandmothers) were cold and distant and showed no pleasure in their daughter's female sexuality and femininity. These women exhibited profound envy and jealousy toward boys; were tomboys until adolescence but eventually married when, with the feminizing of adolescence, "her hopes were wrecked" ever to become a man herself. The fathers were seen as inept and objects of scorn, typically absent either physically or emotionally from the home. The boys were perceived by their mothers as beautiful and each mother engaged in a 'blissful symbiosis' with her 'graceful . . lovely . . cuddly' son, involving shared physical intimacies and boundaries, and physical touching. In a predictable manner the mothers were thrilled to give birth to a son and conferred on him a heroic or masculine name. Having given birth to this 'beautiful . . . phallus', however, she now "has a cure for her lifelong sense of worthlessness." Now began the process out of which developed the boy's femininity. Stoller states: ". . . in the most loving embrace, she keeps her infant against her body and psyche for months and then for years. An excessively close, blissful symbiosis develops . . . it is an endless continuation of the merging of their two bodies from earliest infancy." Consequently, the mother's female gender identity is 'imprinted' on her son. This feminizing process is viewed as "uninterrupted . . . until the boy goes to school." By then it is too late for the boy to change. Indeed, "his femininity fixed, he is not about to change, and so as the years pass, he can only wish to find a way that will allow him to be a girl. He does find that way because now there are techniques for making his body appear female" ([47], pp. 1404–1405).

Stoller's dynamic theory of male transsexualism was so compelling because of its logical consistency that it soon became accepted as the sole explanation of male transsexualism. However, while Stoller takes a conservative view on sex reassignment surgery (SRS), his theory has often been used to support SRS as the treatment of choice for transsexuals. Recently, his theory has been challenged, attacked, and superceded.

Eber [12] has argued that Stoller's views are at best misguided, at worst, false. He points to four flaws in Stoller's theory: (1) pseudolongitudinality; (2) the myth of blissful symbiosis; (3) the misapplication of the concept of imprinting and (4) the assumed immutability of the transsexual's core gender identity. Eber states that there are no published studies linking childhood gender disorders with adult disorders that confirm the adult transsexuals' recall of their early childhood histories. There is one study, however, which disconfirms Stoller's theory [35]. In that study, nine boys were followed from latency through young adulthood. None of those boys with childhood gender identity disturbances became transsexual (though one child toyed with the idea for 6 weeks).

Eber also cited Mahler's [31] work with children to show that Stoller's view that the transsexual enjoys a blissful symbiosis with his mother does not stand up to observational and empirical research with children. Additionally, Stoller's concept of imprinting was also challenged; Eber argued that the ethological term is unacceptable to explain such a complex process as the internalization of psychic structure related to gender identity. Eber also believed that Stoller abandoned his observational viewpoint as a psychoanalyst by removing conflict and trauma from the etiology of the male transsexual's core gender identity (which, incidentally, he did not do for his theory of female transsexualism, which he saw as akin to homosexuality in psychodynamic terms, and rooted in conflict and trauma). Moreover, empirical studies of young children suggest that core gender identity, while anchored around age 2, is an evolutionary concept that changes over time. Eber's objections to Stoller's theory of transsexualism are now widely accepted by most transsexual clinicians. Indeed, Meyer's [33] classic paper on the clinical variants of transsexualism (which focused on the Johns Hopkins transsexual population) failed to identify a group of patients similar to Stoller's 'true' transsexuals.

PSYCHOANALYTICAL THEORIES

While Stoller is a psychoanalyst, his views on transsexualism are not generally accepted by his colleagues. Indeed, one of his colleagues and antagonists, Charles Socarides [44], viewed transsexualism as neither a wish nor a diagnosis. Rather, he viewed it as a perversion, a psychological disorder that was rooted in conflict and trauma, originating sometime during the first three years of development. As a pregenital disorder, and a perversion, Socarides believed that the transsexual's nuclear conflicts forced him into "sexual behavior, which not only affords orgastic release but ensures ego survival." When, as adults, these individuals are stressed, they may regress to wishes for SRS. Socarides believed that the perversion served to repress the transsexual's nuclear conflict; that is, "the urge to regress to a preoedipal fixation in which there is a desire for and a dread of merging with the mother in order to reinstate the primitive mother-child unity" ([44], p. 347). He felt that by achieving femininity through SRS the transsexual was also escaping from

> visible homosexuality, undergoes the dreaded castration . . vicariously identifies with the powerful mother, neutralizes fear of her and consciously enjoys the infantile wish for intercourse with the father (the negative oedipus complex realized); escapes paranoid-like fear of aggression from hostile stronger men who could damage one in homosexual relations ([44], p. 347).

Like Stoller, Socarides believed that it was a pathological family environment that created the fertile soil for the growth of transsexualism. Unlike Stoller, he focused on the role of trauma and conflict in the transsexual child's gender identity development.

Another psychoanalyst, Limentani [23], challenged both Stoller's and Socarides' position. However, he did view transsexualism as a serious psychological disorder arising during the preoedipal period. Some of the etiologic factors in transsexualism included: the father played a prominent role "in the development of persistent transsexual fantasies or serious disorders in gender identity"; there was a lack of introjection of the good breast, "which (was) central to the psychopathology of transsexualism"; the transsexual, while not overtly psychotic, had a major disturbance in symbol formation and his/ her thinking had an "unquestionable psychotic flavor"; "beneath the castration anxiety (lay) the most profound disturbances in object relations and it was separation anxiety (often equalled to a fear of annihilation) which would finally mobilize our attention"; and finally, transsexuals utilized "violent projective identificatory mechanisms."

In sum, the psychoanalytic position is quite diverse. While all three of the above psychoanalysts view transsexualism as arising within the context of a pathological family structure, there is disagreement as to how this takes place. For some, the patient is a target of the family's intense gender conflicts in which trauma defense, ambivalence, and conflict play pivotal roles in the transsexual-to-be's evolving gender identifications. For Stoller, conflict, defense, and trauma play no role in the etiology of 'true' transsexualism (he would view Socarides' and Limentani's patients as not 'true' transsexuals). While the father is also seen as facilitating the child's gender role and identity, it is the role of the mother that seems to be crucial in the etiology of the disorder. The disorder of transsexualism is viewed by most psychoanalysts to be akin to a perversion, interfering with the child's separation and individuation and leading to structural ego defects, impairment in cognition (specifically a defect in symbol formation), and profound impairment in ego functioning. While no one theorist holds all of these ideas it seems as if, with each successive generation of psychoanalysts, the disorder of transsexualism is described as more profound and regressive.

FAMILY DYNAMICS

From a different vantage point, there are a number of studies that have attempted to study the families of transsexuals in depth. In one study, Pauly [37, 38], who focused on the family pathology of transsexuals, concluded that "I can think of few worse fates than to be the victim of the kind of family discord or ignorance which breeds gender identity problems" ([37], p. 522). While this is a rather strong statement, there is certainly reason to believe that in every case (transsexual or otherwise) the parental milieu can have a profound influence on the child's sexual, gender, and self development. For example, Litin *et al.* [24], focusing on more general patterns of sexual development in children, concluded that "unusual sexual behavior evolves by adaptation of the ego to subtle attitudes within the family, a process that distorts the instinctual life of the child." They viewed the child's (and later adult's) perverse sexual acting out as resulting from the parent's "unconscious permission and subtle coercion", a process which leads to a defect in the child's superego and ego.

The transsexual research supports Litin's [24] conclusions and suggests that several factors may be involved, including: the mother's unique role in shaping her child's transsexualism [47]; intergenerational family conflicts, which cause transsexualism in more than one family member [22, 49]; and parental and family dynamics, which shape the child's cross gender identifications, opposite core gender identity, and lead to conflicts in identity formation and instinctual life [8, 52, 53]. At least one author [14] has implicated the role of grandmothers in the etiology of male transsexualism, while others have suggested that the death of a parent has often been the precipitant of the wish for sex reassignment surgery [28]. Newman and Stoller [36] and Bates *et al.* [3] concluded that one ought to intervene with the parents of extremely feminine boys so that the child's gender disorder can be responsive to treatment.

In my book on female-to-male transsexualism [29], I reported on 53 women who requested sex change and discovered that almost half of the patients reported early childhood histories in which child abuse, including violence and neglect, was common and losses, separations, and abandonments the norm. Twelve women reported being sexually abused by family members (in many cases by other women in the family) and two gave birth to their father's child. The family lives of these patients were characterized by disorganization, chaos, and overstimulation, with violence, physical and sexual abuse, incest, and abandonments evident in the majority of families. The pattern was anything but 'stable' (an adjective often associated with the female-to-male transsexual).

In another study on family dynamics and transsexualism, Buck [8] concluded that "in order to assess the critical role played by family dynamics . . . it is most useful to have working data on three generations." Her study indicated that serious gender pathology was found to exist across family generations and that this pathology increased with each successive generation. In summary, it appears that the particular behavior we eventually label as transsexual may be a complex behavioral pattern occasioned by a particular family dynamic.

THE BORDERLINE NARCISSISTIC HYPOTHESIS

In contrast to Stoller's clinical descriptions of the 'true transsexual', most clinicians, following Meyer [33], have noted that there are a wide variety of clinical variants who identify as transsexuals. What these patients seem to have in common (in addition to their transsexual belief) is serious emotional conflict of a preoedipal nature and severe character pathology. Until recently most of the transsexuals' symptoms and pathological character structure were explained as a result of their transsexualism [4]. However, as a result of clinical evidence suggesting that transsexuals have profound psychological problems, a developmental arrest, and primitive mental functioning, many clinicians have speculated about the possible relationship between transsexualism and borderline pathology. Indeed, from a clinical standpoint transsexuals resemble borderlines. Like the borderline patients they appear to have a 'stable instability' to their personality organization and are usually described as quite intact and stable on clinical interview but disorganized and unstable on prolonged evaluation and psychological testing [30]. According to Eber [11, 12] the conceptualization of transsexualism along the lines of borderline pathology also allows the clinician to have more options for treating transsexualism psychotherapeutically. Indeed, the linking of borderline and transsexual pathology has been well established [11, 12, 26, 27, 32, 39, 40, 50, 51].

Person and Ovesey [39, 40] were the first clinicians to link transsexualism to borderline pathology. They viewed transsexualism as having its origins in the preoedipal period and related it to "unresolved separation" anxiety during the separation-individuation phase of development in which the transsexual attempts to "counter separation anxiety . . . [by] a reparative fantasy of symbiotic fusion with the mother." They argued that this fantasy was first conceived before the child was 3 years old. They viewed the male transsexual as

> schizoid-obsessive, socially withdrawn, asexual, unassertive and out of touch with anger . . . [having] a typical borderline syndrome characterized by separation anxiety, empty depression, sense of void, oral dependency, defective self identity, and impaired object relations, with an absence of trust and a fear of intimacy ([39], p. 19).

Volkan [50, 51] also viewed the transsexual as predominantly borderline and believed that the transsexual used primitive splitting as a basic defense in order to maintain the split between his/her male/ female self. Because of a chaotic and defective early childhood milieu the inadequate mothering (parenting) led to a developmental arrest in which the male transsexual needed to rid himself of aggression while the female transsexual needed to become more aggressive. In the former case aggression threatened the destroy the male transsexual's self system from within, while in the latter case a vulnerable female gender-self system was threatened with destruction from without. Volkan viewed the transsexual's request for SRS as really a wish for 'aggression reassignment surgery.'

Investigating a group of transsexuals in West Germany, Meyenburg and Sieguesch [32] also viewed the transsexual's pathology as predominantly borderline in nature. However, many of the transsexual patients in their group, like ones described in my earlier work [27], were affectively unstable and, unlike Person and Ovesey's patients, were impulsive, psychopathic, paranoid, and histrionic. The two different patient groups suggest a broad spectrum of borderline pathology among transsexual patients, reflecting the continuum from lower level (paranoid/ schizoid) to higher level (hysteric/ obsessive) personality functioning [18]. What these patients have in common are the underlying structural criteria for the diagnosis of borderline [18]: that is, lack of identity integration; the employment of primitive psychological defenses; impaired and impoverished object relations; and the capacity for adequate reality testing. The conceptualization of transsexualism as a variant of borderline psychopathology helps us to better understand the structural issues underlying the transsexual's personality organization and enables us to better plan for instituting psychological methods of treatment.

In a recently published book on female-to-male transsexuals [29], I employed the views of Kohut [20, 21] in order better to understand the transsexual's defective narcissism, in which the goals of the transsexual's self system were to be magically transformed into the admired omnipotent cross-gender object in order to bolster their self esteem. Consequently, I see transsexuals as involved in three tasks: attempting to provide stability and cohesion to their precarious gender self-representation; trying to repair their defective ego mechanisms regulating gender-self constancy; and finally, trying to structuralize and consolidate their core gender identity. Essentially, these individuals lack self cohesion and have profound narcissistic self pathology. The 'lost core' to their personalities is an inadequately developed gender-self representation and a lack of an ego mechanism to govern gender-self constancy. During childhood, their mothers failed to empathically relate to, understand, or mirror their children's needs, a process that led to impaired narcissism and an impairment in their ability to fully differentiate self and object images related to gender role and identity. Consequently, these children were left feeling empty and depleted, with little self esteem, an empty depression, and a fixation on the admired object. They rarely achieved a sense of well being and were 'mirror hungry' for external objects whose perceived power and omnipotence were sought after and internalized by transsexuals as part of their weak gender structure. In this sense their major life goal is to restructuralize their self system (that is, strive for a cohesive self system) through a reparative cross-gender fantasy (that is, by becoming the opposite gender and sex they will have a cohesive self system). The reparative fantasy is meant to restore some sense of cohesion to a perceived fragmenting self-system and ward off either a psychosis or personality decompensation. The transsexual fantasy serves both a mirroring and idealizing function: providing the self with a solution to the perceived sense of narcissistic depletion by simultaneously gratifying exhibitionistic, voyeuristic, perfectionistic needs through a fusion with the mirroring and idealizing object. Of course, the solution must fail since it has not been adequately internalized and structuralized as part of the total self system. This may explain why many transsexuals continue to request further surgery in order to achieve a state of perfection (a view that Volkan espouses).

CONCLUSION

Over the last two decades no one has provided clinical evidence supporting Stoller's theory of 'true' transsexualism. In fact, there is increasing clinical evidence that the majority of transsexuals suffer from some kind of borderline pathology. While a few transsexuals may have a biological substrate that organizes their transsexualism, the disorder is primarily psychological. It would appear, therefore, that a psychological disorder deserves to be treated by psychological, not surgical, methods.

The evidence to date suggests that the family member who becomes transsexual evolves his gender pathology through a complex family communication process in which the parents consistently express displeasure and disgust at the child's body image, genitals, and emerging masculinity or femininity. Consequently, these children do not consolidate an appropriate core gender identity and experience severe gender confusion and diffusion. Throughout the preoedipal period the parents assault and thwart their child's development of core gender identity (though to the casual observer none of this might be evident). Reconstructions of childhood experiences through adult eyes suggest that during the *early genital phase* [42] many of the parents overstimulated their 'transsexual' child and provided a chaotic family environment in which the transsexual-to-be child was unable to evolve stable ego mechanisms and functions. Consequently, these transsexual children evolved impaired ego mechanisms regulating gender and self constancy and have a defective self representation. While these children are typically able to separate self and object images (and are not generally psychotic), they are usually unable to differentiate fully their gender-self and gender-object images; have a defect in symbol formation; and evidence a subtle thought disorder. I have hypothesized that during the rapproachement phase of development [29] the parents of these children thwarted any attempt to individuate their appropriate gender self identity. Rather than consolidating an appropriate core gender identity, they evolved gender confusion and diffusion, focusing their feelings of self hatred on their developing body image and ego, experiencing intense hatred over their genitals and disgust at anything related to their sense of maleness or femaleness. I have hypothesized that in order to prevent their gender-self system from complete annihilation they split off their 'all bad' gender images (associated with their appropriate gender) and anchored their identity in an omnipotent 'all good' opposite gender self image, which was highly idealized and supported by their parents. Consequently, these 'transsexuals' experience an oscillation between an 'all good' gender image (associated with the opposite gender) and an 'all bad' gender image (associated with their appropriate gender), which prevents them from ever integrating a normal gender-self system. Moreover, their defensive structure is quite primitive (including such defenses as denial, splitting, projective identification, devaluation, omnipotence). As children they futilely attempted to integrate their fragmented self systems and preserve their very existence by rationalizing that they were 'males or females, trapped in a female's or male's body'. This rationalization provided them with the basis for a pseudo-cohesive self concept and 'repaired' their defective narcissism (while also providing them with a reason for personal survival).

Essentially, the transsexual wish represents the unconscious fantasy for wholeness and integrity of the self system; a defense against annihilation anxiety; an expression of the transsexual's defect in symbol formation; and an overvalued idea that was highly encapsulated and rigidly adhered to (providing a barrier to conscious awareness of and integration of their profound separation/ individuation conflicts and their underlying sado-masochistic fantasies that threatened to engulf and destroy their fragmented self system).

Given the profound nature of the transsexual's self pathology, how can one expect that SRS could ever cure the transsexual's profound psychological conflicts? Indeed, in a major review of the historical, clinical, and bioethical issues of sex reassignment surgery [28] I concluded that psychotherapy, not surgery, was the treatment of choice. All of the research evidence suggests that during the initial 5 year period post surgery the transsexual's psychological problems intensify and become more morbid (although the transsexual may also report being satisfied with the surgery). That is, sexual surgery does not cure psychological problems. Indeed, a number of clinical reports are now suggesting that the critical period is 10 years down the road post surgery, a time frame when a large number of post-operative patients either request reassignment back to their natural gender status or commit suicide. Given these post SRS findings and the newer conceptualizations of transsexualism as a variant of borderline pathology, it is imperative that these findings become integrated into the current legal and bioethical debate in which sex reassignment surgery is often supported as the treatment of choice for transsexuals.

The time has come for clinicians to support vigorously the idea that psychotherapy, not surgery, is the treatment of choice for all transsexuals. While a select few patients may benefit from surgery, they represent a minority of transsexuals (and there are no ways to predict which patients may benefit from surgery). Bioethicists must accept the most current clinical evidence, which suggests that transsexualism is a variant of borderline pathology that should be treated with psychotherapy and not surgery [7, 11, 28, 29, 39, 50].

Institute of Living,
Hartford, Connecticut,
U.S.A.

BIBLIOGRAPHY

1. American Psychiatric Association: 1980, *Diagnostic and Statistical Manual of Mental Disorders (DMS III)*, Washington D.C.
2. Barlow, D. *et al.*: 1973, 'Gender Identity Change in a Transsexual', *Archives of General Psychiatry* **28**, 569–576.
3. Bates, J. *et al.*: 1975, 'Interventions with Families of Gender Disturbed Boys', *American Journal of Orthopsychiatry* **45**, 150–157.
4. Benjamin, H.: 1966, *The Transsexual Phenomenon*, Julian Press, New York.
5. Blumer, D.: 1969, 'Transsexualism; Sexual Dysfunction and Temporal Lobe Disorders', in R. Green, and J. Money (eds.), *Transsexualism and Sex Reassignment*, Johns Hopkins, Baltimore.
6. Brierley, H.: 1979, *Transvestism: A Handbook with Case Studies for Psychologists, Psychiatrists, and Counselors*, Pergamon Press, New York.
7. Brod, T.: 1981, 'The Psychotherapeutic Evaluation of "Transsexuals": Clinical Quandaries Posed by Recognizing Self-deficiencies', in I. Pauly (ed.), *Abstracts and Proceedings of the 7th International Gender Dysphoria Association*, Lake Tahoe, pp. 19–20.
8. Buck, T.: 1977, *Familial Factors Influencing Female Transsexualism*, Unpublished Master's Thesis, Smith College School for Social Work, Northhampton, Massachusetts.
9. Cauldwell, D.: 1949, 'Psychopathia Transsexualis', *Sexology* **16**, 274–280.

10. Davies, B. and Morgenstern, F.: 1960 'A Case of Cycticercosis, Temporal Lobe Epilepsy and Transvestism', *Journal of Neurology, Neurosurgery and Psychiatry* **23**, 247–249.
11. Eber, M.: 1980, 'Gender Identity Conflicts in Male Transsexualism', *Bulletin of the Menninger Clinic* **44**, 31–38.
12. Eber, M.: 1980 'Primary Transsexualism: A Critique of a Theory', *Bulletin of the Menninger Clinic* **46**, 168–182.
13. Ehrhardt, A. and Meyer-Bahlburg, H.: 1981, 'Effects of Prenatal Sex Hormones on Gender Related Behavior', *Science* **211**, 1312–1317.
14. Halle, E., Schmidt, C. and Meyer, J.: 1980, 'The Role of Grandmothers in Transsexualism', *American Journal of Psychiatry* **137**, 497–498.
15. Hill, D.: 1952, 'EEG in Episodic Psychotic and Psychopathic Behavior: Classification of Data', *Electroencephalography and Clinical Neurophysiology* (Amsterdam) **4**, 419–442.
16. Hoenig, J. and Torr, J.: 1964, 'Karyotyping of Transsexualists', *Journal of Psychosomatic Research* **8**, 157–159.
17. Imperato-McGinley, J. *et al.*: 1979, 'Androgens and the Evaluation of Male Gender Identity among Male Pseudo Hermaphrodites with a 5-alpha-reductase Deficiency', *New England Journal of Medicine* **300**, 1233–1237.
18. Kernberg, O.: 1975, *Borderline Conditions and Pathological Narcissism*, Jason Aronson, New York.
19. Kohlberg, L.: 1971, 'A Cognitive Developmental Analysis of Children's Sex-role Concepts and Attitudes', in E. Maccoby (ed.), *The Development of Sex Differences*, Stanford University Press, Stanford, California, pp. 82–173.
20. Kohut, H.: 1971, *The Analysis of the Self*, International Universities Press, New York.
21. Kohut, H.: 1977, *The Restoration of the Self*, International Universities Press, New York.
22. Liakos, A.: 1967, 'Familial Transvestism', *British Journal of Psychiatry* **113**, 49–51.
23. Limentani, A.: 1979, 'The Significance of Transsexualism in Relation to Some Basic Psychoanalytic Concepts', *International Review of Psycho-Analysis* **6**, 139–153.
24. Litin, E. *et al.*: 1956, 'Parental Influence in Unusual Sexual Behavior in Children', *Psychoanalytic Quarterly* **25**, 37–55.
25. Loomis, D.,: 1977, Cognitive Abilities in Male to Female Transsexuals', in *Dissertation Abstracts International*, Microfilms No. 7731000, Ann Arbor, Michigan.
26. Lothstein, L.: 1977, 'Psychotherapy with Patients with Gender Dysphoria Syndromes', *Bulletin of the Menninger Clinic* **41**, 563–582.
27. Lothstein, L.: 1979, 'Group Therapy with Gender-dysphoric Patients', *American Journal of Psychotherapy* **33**, 67–81.
28. Lothstein, L.: 1982, 'Sex Reassignment Surgery: Historical, Bioethical, and Theoretical Issues', *American Journal of Psychiatry* **139**, 417–426.
29. Lothstein, L.: 1983, *Female to Male Transsexualism*, Routledge and Kegan Paul, London.
30. Lothstein, L.: 1984, 'Psychological Testing with Transsexuals: A 30 Year Review', *Journal of Personality Assessment* **48**, 500–507.
31. Mahler, M.: 1967, 'On Human Symbiosis and the Vicissitudes of Individuation', *Journal of the American Psychoanalytic Association* **15**, 740–763.

32. Meyenburg, B. and Sieguesch, V.: 1977, 'Transsexuals in West Germany: Therapeutic Guidelines and Legal Problems', *Abstract: Fifth International Gender Dysphoria Symposium*, Norfolk, Virginia, pp. 1–13.
33. Meyer, J.: 1974, 'Clinical Variants Among Applicants for Sex Reassignment', *Archives of Sexual Behavior* **3**, 527–558.
34. Money, J. and Ehrhardt, A.: 1972, *Man and Woman, Boy and Girl*, Johns Hopkins Press, Baltimore.
35. Money, J. and Russo, A.: 1979, 'Homosexual Outcome of Discordant Gender Identity/Role in Childhood: Longitudinal Follow-up', *Journal of Pediatric Psychology* **4**, 29–41.
36. Newman, L. and Stoller, R.: 1974, 'Non-transsexual Men Who Seek Sex Reassignment', *American Journal of Psychiatry* **131**, 437–441.
37. Pauly, I.: 1974, 'Female Transsexualism: Part I.', *Archives of Sexual Behavior* **3**, 487–507.
38. Pauly, I.: 1974, 'Female Transsexualism: Part II.', *Archives of Sexual Behavior* **3**, 509–525.
39. Person, E. and Ovesey, L.: 1974, 'The Transsexual Syndrome in Males: I. Primary Transsexualism', *American Journal of Psychotherapy* **28**, 4–20.
40. Person, E. and Ovesey, L.: 1974, 'The Transsexual Syndrome in Males: II. Secondary Transsexualism', *American Journal of Psychotherapy* **28**, 174–193.
41. Rekers, G. *et al.*: 1974, 'The Behavioral Treatment of a "Transsexual" Preadolescent Boy', *Journal of Abnormal Child Psychology* **2**, 99–116.
42. Roiphe, E. and Galenson, H.: 1981, *Infantile Origins of Sexual Identity*, International Universities Press, New York.
43. Rubin, R. *et al.*: 1981, 'Postnatal Gonadal Steroid Effects on Human Behavior', *Science* **211**, 1318–1324.
44. Socarides, C.: 1970, 'A Psychoanalytic Study of the Desire for Sexual Transformation ("Transsexualism"): The Plaster-of-Paris Man', *International Journal of Psycho-Analysis* **51**, 341–349.
45. Späte, Z.: 1970, 'Zum Abteil des limbischen Systems in der Pathogenese des Transvestismus', *Psychiatrie, Neurologie, und medizinische Psychologie* **22**, 339–344.
46. Stoller, R.: 1968, *Sex and Gender*, Science House, New York.
47. Stoller, R.: 1975, 'Gender Identity', in A. Freedman *et al.* (eds.), *Comprehensive Textbook of Psychiatry/II*, Williams & Wilkins, New York, pp. 1400–1408.
48. Stoller, R.: 1979, 'A Contribution to the Study of Gender Identity: Follow-up', *International Journal of Psycho-Analysis* **60**, 433–441.
49. Stoller, R. and Baker, H.: 1973, 'Two Male Transsexuals in One Family', *Archives of Sexual Behavior* **2**, 323–328.
50. Volkan, V.: 1976, 'Aggression Among Transsexuals', Paper presented at the *129th Annual Meeting of the American Psychiatric Association*, Miami, Florida.
51. Volkan, V.: 1979, 'Transsexualism: As Examined from the Viewpoint of Internalized Object Relations', in T. Karascu and C. Socarides (eds.), *On Sexuality*, International Universities Press, New York, pp. 189–227.
52. Weitzman, E. *et al.*: 1970, 'Identity Diffusion and the Transsexual Resolution', *Journal of Nervous and Mental Disease* **151**, 295–302.
53. Weitzman, E. *et al.*: 1971, 'Family Dynamics in Male Transsexualism', *Psychosomatic Medicine* **33**, 289–299.

VERN L. BULLOUGH

SEX RESEARCH AND THERAPY

Though it would seem to the casual observer that sex therapy is something peculiar to the last half of the twentieth century, it has existed as long as recorded history. One of the earliest historical records we have deals with the modern problems of impotence. A whole series of surviving clay tablets from the Tigris-Euphrates Valley has been labelled the potency incantations and were to be used primarily with the problems of the flaccid penis, an issue of concern to therapists today.

Some of the techniques advised by the ancients are similar to those used today. Men who had problems gaining an erection were advised to rub their penis, or if possible, have it rubbed by a woman with a special pūru-oil mixed with pulverized, magnetic iron particles, probably to provide additional friction. While their penis was being rubbed, their therapist was to make such statements as "Let his penis be a stick of martu-wood" and request of the man that they were rubbing "Let a horse make love to me" ([3], p. 21). Unfortunately, we do not know how effective such treatment was, but we can surmise it worked for some individuals or the incantations would not have survived.

Sex was important to the ancients, and from the Babylonian period we have clay models of the female sexual parts and stone models of erect penises, a number of terra-cotta models portraying intercourse, and a rather graphic description, Tablet 104 of the Suma Alua, describing sexual activities of human beings in some detail [5]. Though we have a collection of incantations dealing with women, the female incantations seem to deal more with menstrual regularity and not with ability to have orgasm. In fact, there is a vast ancient literature on menstrual irregularity, but the real concern of such literature is a concern with pregnancy, either how to get pregnant or how to encourage a spontaneous abortion or how to avoid pregnancy altogether.

It is not until Roman times that we find any kind of treatment for women that could be described as sex therapy, and the purpose of this therapy was not so much to encourage women to have orgasm as to

E. E. Shelp (ed.), Sexuality and Medicine, Vol. I, 73–85.

avoid hysteria. Galen taught that the uterus had a continual desire to be pregnant, and that unless it was pregnant, at least during the child-bearing years, women suffered from various problems, the only solution to which was either intercourse or pregnancy. This was because, according to Galen, the female secreted a substance similar to the male semen, a substance produced in the uterus, and that could only be expelled by orgasm. Without orgasmic expulsion, the substance would be retained in the body and would begin to spoil, and this spoilage led to a corruption of the blood that in turn led to a cooling of the body and eventually to an irritation of the nerves, which led to hysteria.

Recognizing this, Galen felt it was important for women unable to have regular intercourse to find some other means of releasing this secretion – i.e., through masturbation. He described what happened when one of his woman patients applied warm substances to her genitalia and manipulated herself with her fingers:

> Following the warmth of the remedies and arising from the touch of the genital organs required by the treatment there followed twitchings accompanied at the same time by pain and pleasure after which she emitted turbid and abundant sperm. Thus it seems to me that the retention of sperm impregnated with evil essences had – in causing damage throughout the body – a much greater power than that of the retention of the menses (*De locia affectis*, VI, 2:39).

The ancients also had various kinds of sex manuals, although most of the ones we know about came from India and China rather than the West. The *Kamasutra* of Vatsyayana [35] is one of the earliest of Hindu classics and was believed to have come from the gods. Though some of the positions depicted in the *Kamasutra* are impossible to imitate, it was widely read and copied, and there are hundreds of manuals dealing with positions in Indian, Chinese, and later Islamic cultures [7]. In the West, the closest thing approaching it that has survived is in the writings of Aristotle, and so great was Aristotle's influence that the most widely used sex manual in the English-speaking world from the seventeenth through the nineteenth century was known as *Aristotle's Masterpiece*, although only fragments of it can actually be traced to Aristotle [6].

Aristotle is important in another sense, however, since he could be called the first person to study sex scientifically. Though various individuals before Aristotle had observed enough to know the more superficial aspects of sexuality, Aristotle was the first, or rather the first recorded person, to study sex and reproduction in all its variety. Aristotle taught that all animals reproduced by one of three means: (1) by sexual means,

(2) by asexual means, and (3) by spontaneous generation. In this last category he included a number of lower animals such as fleas, mosquitoes, and flies, which he held were produced out of putrefying substances. In spite of such an erroneous assumption, Aristotle, or perhaps those on whom he relied for information, often proved to be insightful observers of sexual activity. One of his more remarkable observations was the mating of the octopus. He reported that while mating, the two octopi swam about, intertwining mouths and tentacles, until they fitted closely together. Then:

The octopus rests its so-called head against the ground and spreads abroad its tentacles, the other sex fits into the outspreading of these tentacles, and the two sexes then bring their suckers into mutual connection (*History of Animals*, V, 6).

This is a rather accurate description of the mating process in which the male uses his hectostylus (a tentacle serving as the arm of procreation) to remove a semen cartridge from his own mantle and place it in the female's. Because Aristotle's observations were such a mixture of masterful insight and popular superstition, there was a reluctance to challenge his beliefs, even about spontaneous generation, until relatively recent times.

In the medieval period, Aristotelian concepts were supplemented with Arabic notions. Probably the most important addition in terms of sex therapy was the belief in the importance of heavenly movement on one's sex life. At that time, astronomy and astrology were equated one with the other, and one of the tasks that any astronomer undertook was to cast a horoscope. Only after the invention of the telescope did the two split, one to become a science, the other a pseudo science. According to medieval beliefs, the stars were visualized as animate beings of a higher order of humanity and operating with immutable laws. Humans on earth also operated according to immutable laws, and it was hoped that by seeing what was happening in the heavens, one could reach the level of perception here on earth where the soul became one with the heavens. Some writers held that men and women were, in effect, subservient to the stars, with almost everything predetermined. Most Christian writers rejected such deterministic views, with some saying that heavens were only indicators of what already existed, others indicating that the stars gave options on how to act. Still the influence of the stars was accepted, and this traditional belief has become imbedded in our own culture. The term venereal (as in venereal disease), for example, came from the influence of the planet of Venus, which was regarded as especially influential upon female sexual desires, and hence sexual activity came to be regarded as coming under the influence of Venus.

A child born under the influence of Venus would be beautiful and voluptuous, obviously affecting its future sexual life. Medicine in particular came to rely on the stars for indications of possible diagnosis, and women who were planning to become pregnant were urged to record the exact moment of sexual intercourse so that astrological judgment regarding her offspring would be accurate. Most medical writers, while accepting the influence of the heavens, were not adverse to trying to improve sexual enjoyment. Diet was regarded as particularly important. Raw eggs and sweet cheese were believed to stimulate desire, and therapy for sexual disorders usually began with a corrective diet. For example, William of Saliceto, a thirteenth century physician, states that if a man's ability to engage in sexual intercourse is diminished because of general weakness of the body, he should see to it that he chooses foods that cause him to put on weight and that are well tolerated in the first, second, and third digestions. If diminution of coitus is caused by sperm that become too cold and cannot be ejaculated because it has frozen, hot foods should be consumed [27].

Instructions for the proper way to engage in sexual intercourse abound in the medieval medical manuals. Though each author varies somewhat, they all seem to recognize the importance of foreplay, including caressing, kissing, and 'sweet words.' Rubbing of the female genitalia is also recommended until the 'woman's heat level' has been raised enough, whereupon she will signal she is ready for entry. Most of the medical writings pay particular attention to the need for female orgasm so that she can emit her seed, and most counsel the man to save his ejaculation until he knows that the woman has fulfilled her desire.

A number of difficulties with sexual activity are noted by the medical writers, including priapism in both men and women, which is described as an involuntary tension in the sex organs without desire: women experience itching and men have a constant erection. One of the worst disorders, according to William of Saliceto, is that suffered by men who undergo a weakness of the intestinal sphincter immediately after intercourse. A man who is prone to have this involuntary bowel movement in bed tends to have a soft body, experience great desire for a woman, take exceeding pleasure in sexual intercourse, and occasionally soil himself before coitus as well. He should see to it that he uses a suppository before approaching his partner [27].

One of the most widespread of medieval treatises was something known as *De secretis mulierum*, which came to be incorporated into the work known as *Aristotle's Masterpiece*. It was widely believed to have been composed by St. Albertus Magnus in the thirteenth century, but this is a popular belief not backed up by scholarship. I have been able to locate over 100 different printings of the book, almost all of them published anonymously. They were widespread throughout the English-speaking world, and apparently many printers from the seventeenth century to the twentieth who wanted to do something with an idle press printed the work. It is, in a sense, a 'how-to' book, advising clitoral manipulation by the male since 'blowing the coals of these amorous fires' leads to greater satisfaction for both partners.

Most of the emphasis in the manual, however, is on males, since it was assumed that most of the readers were males. In general, both the medieval and early modern manuals emphasize the positive nature of sex. Sex was regarded as a 'gift from God' to be enjoyed. When particular attention was given to women, it was usually concerned with how to get pregnant (or how not to get pregnant) than with any particular non-orgasmic problem. It apparently was assumed that such problems could be cured by a careful and caring male lover who engaged in the proper amount of foreplay [6].

Modern challenges to traditional knowledge about sex began in the eighteenth century, and though we might label this sex research, for the most part it might be called self-fulfilling research, since it was based on theoretical assumptions that, even when they were adopted, were not very well grounded. Perhaps the most influential force in forcing a reassessment of sexual activity was the Lausanne physician S. A. Tissot (1718–97), who taught that all sexual activity was dangerous because it caused blood to rush to the brain, which in turn starved the nerves (making them more susceptible to damage) and thereby increased the likelihood of insanity. In his classic work, *L'Onanisme* (1760) [34], he argued that the worst kind of sexual activity was the solitary orgasm, since it could be indulged in so conveniently and at such a tender age that excess was inevitable. Moreover, since onanism was a Biblical sin, the realization of guilt entailed by such an act opened the nervous system to further damage. Onanism included not only masturbation, but any non-procreative sexual act.

Tissot used some data to draw his conclusions. The period of lassitude following orgasm had been noted throughout history, and in his mind, this indicated the drain on the body system. At the same time, Tissot was very much a person of his own time, and one of the medical theories was that known as Brunonianism, based on the writings of John Brown. Basic to this theory was the notion of excitability, the seat of which was in the nervous system. Too little stimulation was bad, but excessive stimulation was worse because it could lead to debility by exhausting the excitability. Thus, in Brown's mind (and in Tissot's), there were two kinds of diseases, those arising from excessive excitement (sthenic) and those from deficient excitement (asthenic). Mutual contact of the sexes gave an impetuosity to the nerves, a tonic if you will, but intercourse itself could release too much turbulent energy, and if carried to excess, would cause difficulty. Other theoreticians had some variations on the basic theme such as vitalism, which held that disease resulted when the normal life patterns of the body were interrupted. Tissot's own theory was a variant in which he held that disease came from a loss of energy and undue waste, and that wastage eventually caused death. Much of this wastage was natural, i.e., derived from the aging process. Some of the wastage could be restored through nutrition, but even with an adequate diet, the body could waste away through diarrhea, loss of blood, and most importantly for our purposes, seminal emission. Of this wastage, only seminal emission could be controlled by the individual, whether male or female, and though God had intended some wastage in order to create children, it had to be limited to that purpose. Tissot also observed that many of the mentally ill people in institutions continually manipulated their genitals, and he took this as evidence that masturbation caused mental illness. Making Tissot's concepts more acceptable was the confusion of some of the effects of third stage syphilis with an overactive sex life. The third stage of syphilis was not recognized as being due to syphilis until the last half of the nineteenth century and so the tabes dorsalis, the loss of memory, impotence, tremors, heart disease, skin lesions, and other aspects we now know are associated with third stage syphilis were simply equated with too much sexual activity.

Following Tissot, sex research became intermeshed with moral and emotional feelings about sexuality, and the emphasis in much of the so-called scientific literature came to be hostility to sexuality. Even though the males doing much of the writing about sex knew of their own sex drives, they tended to believe that women lacked such drives, some even going so far as to argue that good women only engaged in sex to have children. Women, especially in the last part of the nineteenth century, became subscribers to some of these concepts, in part because it became a weapon of power, in part because they did not want to be pregnant so often. Much of the scientific literature, however, did not spread below the middle class, and even there the rhetoric was often different from reality. A good indication of this is widespread prostitution, and for many males, the prostitute turned out to be an effective therapist [8]. Women just had to suffer in silence.

It was in such a setting that modern sexology began in the last half of the nineteenth century. At first this effort started as a kind of *Diagnostic and Statistical Manual (DSM)* that could be used for physicians and others involved in court cases of patients with sex problems. The classic was Richard von Krafft-Ebing (1840–1902), whose *Psychopathia Sexualis* is still in print [7]. He combined several prevailing nineteenth century theories to explain sexual 'perversion,' (1) the idea that disease was caused by the physical nervous system, (2) the idea that there were often hereditary defects in this system, and (3) the concept of degeneracy. His concept of heredity led him to ask for repeal of some of the harsher penal laws, although his writings also increased the fear of masturbation. Building upon Krafft-Ebing and modifying some of his ideas was a series of others such as Albert Moll [30, 31], Iwan Bloch [4], L. Thoinot [33], Charles Féré [15], and many others [7]. For English-speaking readers, Havelock Ellis (1859–1939) was extremely important. Ellis more than anyone else popularized the concept of the individual and cultural relativism in sex. The result of his studies was published in a monumental series of volumes, *Studies in the Psychology of Sex*, originally issued and revised between 1896 and 1938. Like Krafft-Ebing, Ellis covered most of the variations in sexual behavior, but unlike his predecessor, he exhibited a far more sympathetic understanding of the individuals involved. In a sense, Ellis was a naturalist, observing and collecting information about human sexuality instead of judging it, and as such can be considered the forerunner of the sex researchers of today. Being cautious in his conclusions, Ellis usually avoided any all-encompassing answer. When he turned to questions such as whether homosexuality was inborn or acquired, physical or psychic, he felt there was, perhaps, some truth in all the views. Basically, however, Ellis believed that sexual differences were inborn and nonpathological, although he granted that perhaps there was a higher number of neurotics among deviants than among other groups. Essentially, Ellis' work was a plea for tolerance and for acceptance that deviations from the norm were harmless and occasionally, perhaps, even valuable [14].

Adding to and reinforcing Ellis' work was Magnus Hirschfeld (1868–1935), who was both a homosexual and transvestite [22, 23]. Undoubtedly his own sexual inclinations helped convince Hirschfeld that homosexuality was not a perversion, but his explanation that it was a result of certain inborn characteristics influenced by internal secretions has failed to win many converts. Hirschfeld's value is not so much in theory as in the information he compiled about all forms of sexuality. He founded the first journal devoted to the study of sex and the first Institute of Sexual Science. To assist his research, he gathered an important library of more than 20,000 volumes and 35,000 pictures. He also developed the first widescale study of sex through what he called a 'psychobiological questionnaire' containing some 130 questions that he had some 10,000 men and women fill out. Unfortunately, most of his library, along with the responses to the questionnaires and other information, was destroyed by the Nazi hoodlums shortly after Hitler came to power.

Included in those helping to change attitudes would be Sigmund Freud [16, 17, 18]. Freud agreed with Krafft-Ebing on the necessity of redirecting sexual energies, but where Krafft-Ebing had held that variant sexual behavior came from sexual drives that had been misdirected in their aim or object, Freud held the cause of the misdirection lay in the nervous system and the mind through which the instinctual drive operated. Though Freud actually paid comparatively little attention to most forms of sexual behavior, his followers seized on his concepts to emphasize the environmental and accidental causes of the sexual impulses. Though later behaviorists, stressing learning and conditioning of animals and man, carried this type of environmental and accidental determinism to an extreme, the practical result of both Freudianism and learning psychologies was to suggest that everyone had the potential to channel his drives toward any form of gratification and use any object. This argument undermined the assumption that certain forms of sex were against nature, for nature itself, the instinctual drive, was visualized as being able to express itself in many ways.

Perhaps because Freud accepted sexuality as important, and was willing to discuss it, psychiatrists and psychoanalysts became the dominant sex experts and therapists in the United States. In part this was by default since most American medical professionals did not talk much about sex with their patients, and so when patients came with sex problems ranging from impotence to homosexuality, they were often referred to the psychiatrists. This resulted in many of the sex problems being subsumed into the general category of a psychiatric illness. Just how far this extended was reported by William Masters at the annual meeting of the Society for Scientific Study of Sex in Chicago in 1983.

There Masters told of past treatments for male ejaculatory incompetence. He advised his listeners to make certain that if a case of ejaculatory incompetence appeared, it was important to ask three questions.

First Question: Have you ever been able to ejaculate in another vagina? If the answer is 'no,' you ask the next question. If it's 'yes,' you do not have to bother with the other questions.

Next Question: Have you ever been able to ejaculate with the penis either with manipulation or by manipulation? If the answer is 'no,' you have to ask the last question, which is, have you ever had a wet dream?

If it's no to all three things, before you ever do psychotherapy, you must introduce this individual to a competent – and I mean a really competent – urologist . . . to find out whether this man has congenital absence of the ejaculatory ducts.

Masters reported that of the six men he had seen who said no to all three questions, three had congenital absence of the ejaculatory ducts and these three had a combined total of 23 years of psychotherapy for ejaculatory dysfunction. "I find that inexcusable. I find that malpractice. All you have to do is ask three questions." Unfortunately, psychiatrists usually did not ask such questions. Instead, they relied on psychoanalytic theory to explain the causes of homosexuality, of transvestism, and of other forms of sexual behavior. Much of this theory was based on untested and unproven assumptions. Undoubtedly the psychiatrists also helped a great many people accept themselves, and simply being willing to listen to people talk about their problems was important. Unfortunately, many went beyond their own expertise and tried to fit various forms of sexual behavior into broad categories to be explained by domineering mothers or absent fathers or other kinds of similar generalizations that lacked empirical verification.

Fortunately there was also another tradition, and there were always a few Americans, mostly physicians, studying sex in a nonpsychiatric model. Perhaps the most important of these was Robert Latou Dickinson [15], whose earliest case studies go back to the 1890s. Others who were important include Clelia Mosher [13, 32], Katherine B. Davis [11], and G. V. Hamilton [20]. There were many others [7], including physicians such as George W. Corner, who later ended up with the

Rockefeller Foundation. It was through his influence that Kinsey was funded in the 1940s.

Dickinson, along with W. F. Robie and LeMon Clark, was responsible for the introduction into American gynecological practice of the electrical vibrator or massager, a device producing intense erotic stimulation and even orgasm in some women who previously had been unable to reach a climax. It was Dickinson's theory, as well as his collaborators', that a woman who had once achieved orgasm, even with a vibrator applied to the clitoris, was more likely to proceed to orgasm during coitus or through masturbation. Dickinson used a glass tube resembling an erect penis in size and shape to observe the behavior of the vaginal lining and cervix during orgasm. In the process, he proved once and for all that women did have orgasms involving physiological changes.

Kinsey represented a watershed in sex research. Much of the earlier research in the aftermath of the Nazi takeover was lost or forgotten. In the meantime, the leadership in science, both social and natural, had been assumed by the United States. Thus, when the Kinsey studies were published, they not only had domestic impact, but worldwide. The Kinsey studies were based on the detailed sexual behavior of 12,000 Americans of both sex and all ages, unmarried, married, and formerly married, drawn from every state and from every educational and socioeconomic status. Kinsey originally had planned to interview some 100,000 individuals, but death intervened, and his successors at the institute he founded have since chosen other paths. In terms of sample statistics, 12,000 individuals ought to have given an effective indication of American sexual behavior, but Kinsey's sampling techniques have often been criticized by social scientists because his interviewees tended to be self-selected or to belong to groups to which Kinsey had an entree. But even those who criticized his sampling techniques have been impressed by the size of his sample and the results he obtained [24, 25].

Kinsey examined sex from the point of view of a social scientist, even though he was a biologist, and his purpose was to find what kind of sexual activities people engaged in, not to condemn or even to define what was natural or unnatural. In this sense, he was building on the foundation laid by Havelock Ellis, Magnus Hirschfeld, and others. Kinsey also did some of the same type of physiological research that Dickinson had started, but the persons most responsible for emphasizing the physiological basis of sexual activity was the team of William H.

Masters and Virginia Johnson. Masters had begun his study with prostitutes (both female and male), but soon found he was able to extend it to non-prostitutes. For his initial study of 382 women and 312 men, a total of 694 individuals participated and all told, they experienced a total of 10,000 orgasms under laboratory conditions. Volunteers masturbated either with their hands or with mechanical vibrators, had artificial coition with a transparent probe similar to that used by Dickinson except that it was electronically controlled, and engaged in various types of sexual intercourse with each other. In the process Masters, and Johnson, who soon had become his collaborator, were able to answer many of the previously unanswered questions associated with the physiology of the sex act, particularly in terms of the female response cycle. Masters and Johnson also developed ways to bring effective help to men suffering from premature ejaculation or impotence and to anorgasmic women [28, 29]. Others, building on Masters and Johnson, developed their own techniques for therapy, as Marilyn Fithian and William Hartmann did [21].

In the post Kinsey era, sex research began to be pushed by various groups, but particularly the Society for the Scientific Study of Sex, which picked up where the pre-World War II journals left off. The Society was soon joined by other journals such as the *Archives* and by an outpouring of research into standard professional journals. Therapists joined in AASECT, and other groups' public perceptions about sex research began to change, aided by sex educational organizations such as SIECUS. As of this writing, sex research seems to have become a legitimate occupation, and sexual therapy is regarded as an important specialty.

State University College of Buffalo,
Buffalo, New York,
U.S.A.

BIBLIOGRAPHY

1. Aberle, S. D. and Corner, G. W.: 1953, *Twenty-Five Years of Sex Research*, W. B. Saunders, Philadelphia.
2. Aristotle: 1910, 'History of Animals', in J. A. Smith and W. D. Ross (eds.), *The Works of Aristotle*, trans. D. W. Thompson, vol. 4, Clarendon Press, Oxford.
3. Biggs, R. D.: 1967, SA.ZI.GA: *Ancient Mesopotamian Potency Incantations*, in *Texts from Cuneiform Sources*, II.

4. Bloch, I.: n.d., *The Sexual Life of Our Time*, trans. from 6th German edition by M. E. Paul, Allied Book Company, New York.
5. Boissier, A.: 1893, 'Summa Alua', *Revue Sémitique* **1**, 171 ff.
6. Bullough, V. L.: 1973, 'An Early American Sex Manual, or Aristotle Who?', *Early American Literature* **VII**, 236–247.
7. Bullough, V. L.: 1980, *Sexual Variance in Society and History*, University of Chicago Press, Chicago.
8. Bullough, V. L. and Bullough, B.: 1978, *Prostitution: An Illustrated Social History*, Crown Publishers, New York.
9. Bullough, V. L. and Brundage, J.: 1982, *Sexual Practices and the Medieval Church*, Prometheus Books, Buffalo.
10. Cresbron, H.: 1909, *Histoire Critique de l'Hystérie*, Asslin Houzean, Paris.
11. Davis, K. B.: 1929, *Factors in the Sex Life of Twenty-Two Hundred Women*, Harper, New York.
12. Degler, C. N.: 1974, 'What Ought to Be and What Was: Women's Sexuality in the Nineteenth Century', *American Historical Review* **79**, 1467–1490.
13. Dickinson, R. L. and Beam, L.: 1932, *A Thousand Marriages*, Williams & Wilkins, Baltimore.
14. Ellis, H.: 1936, *Studies in the Psychology of Sex*, Random House, New York.
15. Féré, C.: 1932, *Sexual Degeneration in Mankind and in Animals*, trans. U. van der Horst, Anthropological Press, New York.
16. Freud, S.: 1910, *Three Contributions to Sexual Theory*, Journal of Nervous and Mental Diseases Publishing Company, New York.
17. Freud, S.: 1924–50, *Collected Papers*, Imago Publishers, London.
18. Freud, S.: 1938, *Basic Writings*, Modern Library, New York.
19. Galen: 1586, *De locia affectis*, VI, Venice.
20. Hamilton, G. V.: 1929, *A Research in Marriage*, A+C Boni, New York.
21. Hartman, W. E. and Fithian, M.: 1974, *Treatment of Sexual Dysfunction*, Jason Aaronson, New York.
22. Hirschfeld, M.: 1920, *Die Homosexualität des Mannes und des Weibes*, 2nd ed., Louis Marcus, Berlin.
23. Hirschfeld, M.: 1910, *Die Transvestiten*, Alfred Pulvermachier, Berlin.
24. Kinsey, A. *et al.*: 1953, *Sexual Behavior in the Human Female*, W. B. Saunders, Philadelphia.
25. Kinsey, A. *et al.*: 1948, *Sexual Behavior in the Human Male*, W. B. Saunders, Philadelphia.
26. Krafft-Ebing, R. von: 1894, *Psychopathia Sexualis*, trans. from the 7th enlarged and revised German by C. Gilbert, F. A. Davis, Philadelphia.
27. Lemay, H. R.: 1982, 'Human Sexuality in Twelfth- through Fifteenth-Century Scientific Writings', in V. L. Bullough and J. Brundage (eds.), *Sexual Practices and the Medieval Church*, Prometheus Books, Buffalo, New York, pp. 187–205.
28. Masters, W. E. and Johnson, V.: 1966, *Human Sexual Response*, Little Brown, Boston.
29. Masters, W. E. and Johnson, V.: 1970, *Human Sexual Inadequacy*, Little Brown, Boston.
30. Moll, A.: 1913, *The Sexual Life of the Child*, MacMillan, New York.

31. Moll, A.: 1931, *Perversions of the Sexual Instinct*, Julian Press, Newark, New Jersey.
32. Mosher, C.: 1923, *Woman's Physical Freedom*, The Woman's Press, New York.
33. Thoinot, L.: 1911, *Medicolegal Aspects of Moral Offenses*, trans. from the French by A. W. Weysse, F. A. Davis, Philadelphia.
34. Tissot, S. A.: 1769, *L'Onanisme, Dissertation sur les Maladies Produits par la Masturbation*, 4th ed., Lausanne.
35. Vatsyayana: 1962, *Kama Sutra*, trans. R. Burton, Dutton, New York.

FRITZ K. BELLER

A SURVEY OF HUMAN REPRODUCTION, INFERTILITY THERAPY, FERTILITY CONTROL AND ETHICAL CONSEQUENCES

Modern development in reproductive medicine started with fertility control. The introduction of 'the pill' by Pincus and Rock in 1958 stimulated research in this area in both human and animal species. However, it is felt by many that biologic reproductive technology has reached uncontrollability. The situation has been compared to a similar development of the atomic bomb some decades ago. This essay is a brief review of reproductive biology, infertility, contraception, and selected related ethical issues.

SEX DEVELOPMENT

Many societies consider life to begin not at a particular stage of intrauterine development but with the emergence of the newborn baby. This belief is questioned by many, especially since newer research has demonstrated that the fetus is capable of communication with his environment. The fetus can demonstrate the sensation of pain as well as other sensations.

Females

Pubertal changes in females are first observed at about 8 years (range 7–10 years) and end with the first episode of menstrual bleeding, which is termed the 'menarche'. Subsequent menstruation may occur for up to two years without ovulation. The normal menstrual cycle in the female, whereby the lining of the endometrium is more or less completely shed every four weeks with consequent bleeding, depends on a very complex endocrine system.

The hormonal function of the ovary is intimately dependent on the presence of one responsive follicle. There are actually three stations of hormonal production that interact by a positive or negative feedback. Two production sites are located in the brain. The hypothalamus se-

E. E. Shelp (ed.), Sexuality and Medicine, Vol. I, 87–108.

cretes a decapeptide that stimulates the release of the 'gonadotropins' and thus is termed 'gonadotropin releasing hormone' (Gn-RH). The term 'luteinizing releasing hormone' (LH/RH) is used synonymously. The anterior lobe of the pituitary gland secretes two gonadotropic hormones, the follicle-stimulating hormone (FSH) and luteinizing hormone (LH). The ovary is the target gland that produces a variety of hormones, the most important ones being estrogens, progesterone and, to a lesser degree, androgens. The hypothalamus, pituitary gland, and ovaries are part of a complex self-control system that operates by a so-called feedback mechanism. Stimulating effects of gonadal steroids on gonadotropin secretion are termed 'positive feedback'; inhibiting effects are termed 'negative feedback'.

The hypothalamus is functionally linked to the anterior lobe of the pituitary by means of a venous capillary system – the portal circulation. The hypothalamic releasing hormone Gn-RH is secreted into this portal circulation in small intermittent spikes. These induce synchronous discharges of LH and less prominent discharges of FSH from the pituitary gonadotrophs. While this pulsatile secretion of the gonadotropins stimulates follicular growth, the latter produces increasing amounts of estradiol. This steroid feeds back with gonadotropin release: throughout the major part of the follicular phase an inhibitory effect increases the storage of mainly LH in the pituitary and at midcycle by a positive effect this gives way to a precipitous release of the stored material. The latter mechanism results in the midcycle LH-peak, which in turn is necessary for ovulation.

After the egg is expelled the ruptured follicle is transformed into a corpus luteum (yellow body), which now produces considerable amounts of another steroid – progesterone. This hormone reduces hypothalamic Gn-RH release by decreasing the frequency of Gn-RH pulses. Consequently, gonadotropin pulses are released less frequently from the pituitary and a second LH peak or further follicle growth is avoided. Usually progesterone secretion lasts for 14 days and during this time gonadotropin levels in the peripheral circulation are rather low.

If the ovum has not been fertilized, a complex mechanism including ovarian steroids, prostaglandins, and as yet unknown small peptides is started within the ovary at about the mid-luteal phase, leading to the spontaneous breakdown of the corpus luteum approximately one week later. The drop in progesterone levels and the withdrawal of estradiol originating from the corpus luteum also produces, by a complicated scheme, ischemia in the endometrium and bleeding. However, the falling levels of ovarian hormones gradually reduce the inhibition of gonadotropin and release hormone secretion, whereby these start to rise again and a new cycle begins (cf. [6]).

Between the ages of 40 and 50 the ovary becomes less sensitive to the gonadotropic hormones. Ovulation occurs less frequently and finally ceases. The menstrual cycle becomes monophasic. Around the age of 50 to 55, bleeding episodes cease. This period is called menopause and represents a termination of the immediately preceding years, which are known as the climacteric or preclimacteric phase. During the next ten years, due to insufficient feedback (lack of estrogens), FSH is very high and as it slowly decreases the senium is reached.

Males

It is suspected that spontaneous or drug-induced changes in the hormonal balance of the mother during early pregnancy may influence sexual orientation and behavior of the child in adulthood, perhaps accounting for at least some cases of homosexuality (cf. [10]). Sexual characteristics in the male are caused by testosterone secretion, which is the result of maturity in the testes. Seminiferous tubules occupy most of the testicular volume. During development the seminiferous cords, present in the fetal testes, increase in length, producing the seminiferous tubules. They contain germinal cells producing spermatozoa.

Testosterone is produced by Leydig cells, which are located in the testicular interstitium and may be metabolized to other steroids (dihydrotestosterone and estradiol). These cells are stimulated by luteinizing hormone (LH), secreted by the pituitary. Under the influence of testosterone the penis grows in length and width. Pigmentation and wrinkling (rugosa) of the scrotal skin also occur. The internal sex organs (seminal vesicles, prostate gland, and bulbo-urethral glands) enlarge, accompanied by an increase in masculinity, e.g., deepening of the voice, masculine hair pattern, etc.

In the male there is also a hormonal feedback mechanism. FSH stimulates spermatogenesis and LH testosterone secretion. The function of the testes prior to puberty is unclear. Nocturnal LH pulses occur during puberty and induce increased testosterone production. In contrast to previous views that testosterone levels decrease with age, newer data reveal normal endocrine and exocrine testicular function in healthy older men (cf. [9]). The spermatogenetic cycle in men takes 74 ± 6 days. All stages of spermatogenesis can be observed simultaneously in the same seminiferous tubule. Spermatozoa are released into the lumen and transported to the epididymis where they are stored and mature further. Seminal plasma is secreted by the seminal vesicles and the prostate gland.

HUMAN DEVELOPMENT

At intercourse about 300,000,000 sperm are deposited in the vagina. They move upwards through the cervical canal, into the uterine cavity, and reach from there the fallopian tubes. In general, one ovum out of approximately 300,000 present in the ovary develops; it is called the Graafian follicle. Ovulation, which occurs around the midpoint of a 28-day menstrual cycle, can be retrospectively determined by subtracting 14 days from the first day of the following menstruation. The second fortnight after ovulation is stable and irregularities of the menstrual cycle are due predominantly to irregularities of the pre-ovulatory phase. The fallopian tube picks up the egg from the ovarian surface or fishes it out of the fluid present in the cul de sac. The egg, now located in the ampulla region of the fallopian tube, is penetrated by just one sperm. The sperm contains several enzymes located on the acrosomal membranes, among them hyaluronidase, acrosin, and a corona-dispersing enzyme. Fertilization results potentially in a new human life. New genetic information is brought by a characteristic DNA structure from the father through the sperm and from the mother through the ovum, producing a genetic identity. The two gametes possess a haploid number of chromosomes at the time of fertilization when they have undergone meiotic (reduction) divisions. The ovum has undergone one complete meiotic division and is in the metaphase stage of the second meiotic division at the time of ovulation. At the completion of each meiotic division of the female gamete, a polar body is extruded. At ovulation the ovum has a polar body that is not extruded until after penetration of the ovum by a spermatozoon. Penetration activates the completion of the second meiotic division. This zygote moves through the fallopian tube in the direction of the uterus while being multiplied by 2, 4, 6, 8, 16, etc., at a rate of almost one division per day.

After six or seven days the blastocyst, as it is called after the first week, reaches the glandular lining of the womb and the process of implantation begins. This process is very crucial. It is of great significance from an ethical point of view. One pole of the sphere of cells, the trophoblast (later to become the placenta), burrows itself into the glandular lining; the opposite pole of this sphere will become the fetus.

The early forming placenta produces hormones that enter the maternal bloodstream and prevent further menstruation by preventing the catastrophic event of early expulsion of the implanted egg together with the endometrium lining. Since the time interval between ovulation and menstruation is only 14 days, and the first 7 days have been passed in the tube, there are only 7 days left for hormone production that will prevent menstruation and sloughing off of new life with the glandular lining. These hormones (human chorionic gonadotrophin or HCG) allow the diagnosis of pregnancy by immunologic techniques.

The implantation process ends exactly 14 days after fertilization. From now on this human life has a biological identity. It is now called an embryo, a status that ends four weeks after fertilization. The German embryologist Blechschmidt has written extensively on the meaning of individualization. It is of great significance that until the end of implantation the blastocyst can split into identical parts and thereby form identical twins. However, the opposite can also occur and twins and triplets can be recombined into a single individual. Such an individual is called a chimera. Therefore, up to the end of the implantation process the blastocyst is not as yet irreversibly an individual, since it still may be recombined with others into one new fetal being (cf. [7]). The chimera has the genetic type XX-XY, which indicates that they are recombinations into one human being of the products of more than one fertilization. These individuals are gonadally disturbed, consisting as they do of a genetic mixture of male and female, and can contain two distinct populations of red blood cells and may have heterochromia of the eyes. These observations are of great ethical significance.

After 4 weeks, i.e., 2 weeks post implantation, the heart begins to function, and another fortnight later (6 weeks after fertilization) all of the internal organs of the fetus are present. Electrical activity in the brain is already present at the end of 8 weeks. Fingers and toes are fully recognizable. After 12 weeks the brain structure is complete. The fetus has recognition ability and begins to communicate with its environment. The development process continues until 40 weeks (280 to 281 days after conception or 267 days after ovulation), when spontaneous labor begins and the fetus is born.

CONTRACEPTION

For centuries the most popular method of contraception has been coitus interruptus, supported by abortion, which was for a time seen by Western societies as criminal. Hellegers indicated that the number of births declined steadily in the USA and, by 1921, had reached zero population growth, meaning that the average couple had 2.3 children, which is generally sufficient to replace both parents (cf. [7]). This is known to have happened in France in the mid-nineteenth century, long before the discovery of the vulcanization of rubber.

Another contraceptive technique, implemented through altering social standards, is child spacing and child packing. China is an excellent example of this. The earlier in life women bear their children, the more offspring they tend to have. According to one calculus, if the first child is born at the age of 30 rather than 20, the number of offspring in a decade drops from 5,000 to 300. This is the basis for China's law setting marriage age for men at 26 and for women at 24. The 'One Child Marriage' is a more effective way of decreasing population in developing societies than modern contraceptive techniques that are successful in the industrialized world. Two factors help explain this: (a) lack of motivation, and (b) lack of resources. Interestingly enough, religion does not play a significant role in the developing countries. If infant mortality in India is 50% and religion requires that a father's burial by his own son is necessary to reach eternity, then it is easy to understand that at least 10 children are required to meet that goal. It is therefore of little consequence that Hinduism does not restrict contraception. For some time it was believed that Taiwan would provide an excellent example of the effectiveness of modern contraceptive techniques, until it was realized that the increasing use of contraceptives simply parallels industrialization.

The effectiveness of a given contraceptive method has been expressed since 1936 by the Pearl Index. This is a theoretical rather than a factual indicator of effectiveness. The Pearl Index expresses effectiveness by the number of pregnancies per hundred 'women-years' of use, or precisely 1200 months of total exposure. The problem with this index is that controls were rarely used and the pregnancy rate of 1200 unprotected cycles is not fully 100%, but rather about 40%. The Pearl Index is therefore not expressible on a percentage basis. Since no allowance is made for dropouts, it is understandable that the numbers vary greatly from table to table, given by various authors, and can only be considered semiscientific at best. Modern statistical methods, e.g., the life table method, are currently being used to replace the Pearl Index (cf. [13]).

Temperature Methods

These are mostly based on the previously mentioned fact that the phase of the cycle, after ovulation, is stable for 14 days. Another requirement for this method is knowledge that the ovum may be fertilizable for 24 to 48 hours after ovulation. Sperm may be capable of fertilization much longer than previously thought.

Ogino and Knaus developed a technique requiring that a woman measures the duration of her cycles for at least a year and then use the mean. This is called the calendar or rhythm technique. The method was improved by measuring basal body temperature and charting results. By this method, an increase in temperature of one degree centigrade indicates ovulation (temperature rhythm). Both techniques may result in failure when, due to minor psychological impacts (e.g., traveling) ovulation is delayed. Billings' method is based on the change of cervical secretion during the menstrual cycle. An improvement of this technique results from more accurate measurements of vaginal secretions.

Coital Techniques

Coitus interruptus requires withdrawal of the penis from the vagina prior to ejaculation. It is presumed to be the most frequently used technique in the world today. This has not changed from previous centuries. Efficiency of this method is considered low. Some authors feel that even the first few drops of seminal plasma, being excreted before ejaculation, contain a high concentration of sperm. This has been contradicted by others who have observed high concentrations of sperm only in the second of a six-portion ejaculation. Non-genital coitus, e.g., penile anal coitus, fellatio, cunnilingus, mutual oral-genital coitus, and mutual masturbation, are practiced by some for contraceptive purposes.

Barrier Methods

Condoms: This type of contraceptive device was originally made from linen or animal intestine. Today they are made of rubber or synthetic material with or without a spermicidal agent as coating. Manufacturers' quality control specifications accept a breakage rate of 2.5 : 1000 condoms.

Diaphragm: This device was developed by Mensigma in Europe but was never popular there. In the USA, through the efforts of Margaret Sänger in New York, it became a success. The device consists of a circular spring covered by a stretched piece of rubber or synthetic material. Sizing is determined by a physician's measurement. After pregnancy, regardless of whether ending in an abortion or a full-term delivery, the size must be determined. The diaphragm is inserted before coitus and should be left in place for at least 6 to 8 hours after the act. It may be covered by a cream for easier insertion, and this cream may contain a spermicide.

Vaginal Spermicides: The active substance in such agents is a modification of nonoxynol-9, which is provided in a variety of vehicles and available as creams, suppositories, or foams. Reapplication is necessary for each episode of coitus. Vaginal spermicides are presently undergoing improvement. Substances are inserted into the vagina via various new application techniques.

Intrauterine Devices: It is said that Bedouins used pebbles placed in the uterus of their camels to prevent pregnancy. In 1928 the German physician Gräfenberg developed his ring, a device which was made of silver, which later was determined to be of an alloy since so-called 'German silver' after World War I was actually 60 percent copper. Forty years later, Wagner showed that the 'Gräfenberg Ring' liberated as much copper as the modern copper-releasing devices. Gräfenberg had demonstrated through more than 12,000 carefully observed cycles a very small infection rate. Contrary reports in the literature failed to acknowledge the historical fact that the ring was not differentiated from cervical spreading devices which had a very high infection rate. The Gräfenberg device was banned in 1933, and its inventor was imprisoned and exiled, in 1940, to the United States. In the meantime the Japanese Obio had tried plastic devices.

Shortly after Gräfenberg's death in 1958 a resurgence of the intrauterine device (IUD) began. Nearly all geometrical forms, made from various materials, preferably plastics, were attempted; only a few proved effective. The most popular and frequently used all over the world is the Lippes Loop. Another popular device, the Dalcon shield, turned out later to be associated with a high rate of infection. Tatum attempted a plastic device in the shape of a T, which was, however, very ineffective. On the basis of data of Zipper, who had found that metallic ions were effective as contraceptives, IUDs came to be fitted with a copper wire around the stem. Due to their release of ions, they came to be called releasing devices (cf. [12]). Another T-shaped device releases progesterone. The problem with this device is that the amount of hormone that can be incorporated lasts only about one year, after which the device must be replaced. The T itself has proven a failure (cf. [2]).

Different types of IUD can stay in the uterus for different lengths of time. The Lippes device can stay in place for ten years and longer; copper-releasing devices for up to five years; and the progesterone-releasing devices for one year. The mode of action of the IUD is unknown. Sperm migration may be inhibited, as well as implantation. While plastic devices need a certain minimum surface area, copper-releasing devices can be effective in smaller sizes. The effectiveness is improved if the device is located high up in the fundus. This can be demonstrated by ultrasonic examination. The copper T has a Pearl Index pregnancy rate of 0.6, as measured over 15,000 cycles by one investigator who examined the device at least once a year by ultrasonography. According to Lehfeldt the device prevents intrauterine pregnancy in 95% of cases, 90% of ectopic pregnancies, and does not affect ovarian pregnancies at all. The patient failure rate is very low (cf. [2]).

Most frequently observed complications of IUDs are hypermenorrhea, prolonged menstruation, and spotting. The origin of these is unknown. Infection is less frequent but more serious because of associated problems of infertility. A higher rate of infection is associated with promiscuity, lack of personal hygiene, and gonorrheal infection. The perforation rate varies with the experience of the introducer, from 1 : 200 to 1 : 20,000.

Oral Contraception

Oral contraceptives are female hormones taken singularly or in combination over long periods of time. The first 'pill' was developed by Pincus and Rock, who used a high dose combination of estrogens and progesterone. Over the last 20 years the dosage has been reduced, since it has been shown that effectiveness remains unaltered and side-effects are minimized through continued decrease of hormone administration.

At present oral contraceptives consist of either of two estrogens, ethinyl estradiol or mestranol, the latter being a derivative of the former. The seven synthetic progestins used for progesterone activity are: norethynodrele, ethynodiol diacetate, norethindrone, norethindrone acetate, lynestrol, norgestrel, and desogestrel, all derivatives of the 19-carbon androstane nucleus. Although they are not natural derivatives they do have progesterone activity. (Progesterone itself is effective only in very high doses given parenterally.) They are termed progestins or progestagens or gestagens, and they can to some extent be metabolized to estrogens. There are considerable difficulties in estimating dose effects in the presence of estrogen.

There are three schemes for administration:
(1) the combination pill;
(2) the sequential pill; and
(3) the minipill.
The combination pill consists of one of the estrogens in a dose of between 30 μg to 50 μg, together with one of the progestagens. The sequential pill is designed to simulate the hormonal level of a normal cycle. The pills for the first ten days contain only estrogens; the remaining pill, estrogen and progesterone. Finally, the minipill consists of low-dose progestagen. The contraceptive effect of the minipill is primarily interference with cervical mucus, which blocks sperm penetration. It is believed that the small amount absorbed into the circulation is without systemic effect. The combination and sequential pills inhibit ovulation and alter the endometrium to prevent implantation. The pill has a low drug failure but a high patient failure rate, the latter being highest with the minipill because this variety has to be taken at a given time± 2 hours every morning. When the pill is taken regularly, prevention of pregnancy is nearly 100%.

With the exception of the minipill, which is taken continuously, the combination or the sequential pill is taken from the fifth day after the onset of the menstrual cycle and continued for 21 days. Two to three days later the next menstruation usually begins and is in amount less and shorter in duration. Some pills consist of a 28-day pack wherein 7 are inert. These pills are also taken continuously.

Minor untoward effects are nausea, vomiting, breast tenderness, and weight gain. These rarely cause termination of pill use. Break-through bleeding is related to low estrogen content and therefore frequently experienced by users of the minipill. Serious side reactions are thromboembolic complications, e.g., thromboembolism, thrombotic stroke, and myocardial infarction. These complications are dependent on duration of intake, age and other risk factors, especially smoking. These side-effects are estrogen-related and very infrequent when the estradiol dose remains below 50 μg. The progestagen component may, in rare instances, increase blood pressure, an effect which is reversible if the pill is stopped. Benign liver tumors (hepatomas) are rare but may be serious because of severe hemorrhage from a liver rupture. The tumors regress after cessation of the pill but may recur during a subsequent pregnancy. Serum lipid changes in the blood are due to a high progestagen content and its residual androgenic action. The only link between the pill and cancer at the present time is the possibility that combination pills may protect against endometrial cancer. Claimed associations between the pill and breast cancer have not been proven.

The 'morning after pill' is a principle rather than a product. It consists of oral applications of estrogens in high doses for 5 days after intercourse which makes the endometrium unfit for implantation. After termination of the hormone, shedding of the endometrium takes the implanted egg with it, if it has implanted. Another technique is to insert an IUD a few hours after intercourse. The morning-after principle can be abortifacient but it is not necessarily so.

Sterilization

Sterilization means inducing permanent means to prevent pregnancy. This can be done in the female or the male. Surgical techniques and anesthesia have been refined to permit a safe, rapid procedure with limited postoperative periods. Laws in most countries permit voluntary sterilization so that the distinction between medical, obstetrical, or social indication for the procedure is no longer necessary.

ABORTION

In many countries a liberalized abortion law has replaced restrictive older laws. In most Western countries human life is protected by law. In some countries abortion is legal up to 12 weeks of gestation (Austria, France, Sweden), in others until 24 weeks (USA). The difference is explained by opinion. Abortion until 12 weeks of gestation is based on methodology, since before the twelfth week the pregnancy can be terminated by suction curettage. This is not possible after the twelfth week. From this time on the protective mechanism provided by nature becomes so effective that various techniques have to destroy the fetus, which is then born prematurely. Usually substances are instilled into the amniotic fluid, such as hypertonic salt and others, which induce endogenously prostaglandin activity, which in turn induces labor. This can also be simulated directly by introducing certain prostaglandin derivatives. These techniques are associated with a considerably higher complication rate than suction curettage before the twelfth week of gestation. The 24-week limitation is identical with the limitation involved in the term 'spontaneous abortion' as defined by the WHO. The weight of the fetus is at this time approximately 500 g. Below this size the fetus cannot sustain life despite more months in utero. Hysterectomy to remove the fetus is indicated only rarely (e.g., vaginal hysterectomy until 14 weeks of gestation or radical abdominal hysterectomy for cancer).

Voluntary abortion should never be a contraceptive technique. However, when used to back up contraceptive failure it may substitute for less safe techniques, thereby preventing complications (e.g., sepsis) and reducing mortality rates (cf. [13]). Voluntary abortion is of course a topic of strong ethical debate.

INFERTILITY

The term 'infertility' is used when a couple had failed to achieve pregnancy after one year of unprotected coitus. The term 'primary infertility' is used when no pregnancy has occurred, and the term 'secondary infertility' is used when previous pregnancies have occurred, resulting in either birth or spontaneous abortion. The term 'infertility in the female' is also used to define the inability to carry a pregnancy to the stage of fetal viability outside the uterus, which is approximately the 24th week, and includes the repeated loss of viable infants in the perinatal or neonatal period. The term 'sterility' is attached to the failure to become pregnant. It is estimated that approximately 10% of couples in industrial countries are infertile (60% of which is accounted for by the female and 40% by the male). In females approximately 50% of these cases are organic and 50% are functional in origin.

In order to become pregnant, sexual intercourse at ovulation is mandatory. The lifespan of the ovum is about 48 hours, that of spermatozoa about 24 hours. If the menstrual cycle is short and menstrual bleeding extended, ovulation may occur during menses. In this case, if the couple objects to intercourse at bleeding, pregnancy is impossible. If a woman has irregular menstrual cycles she may never be aware of the time of ovulation. Another reason she may be unable to become pregnant may be the frequent or regular absences of the husband due to his profession.

Functional Defects in the Female

Absence of ovulation can result from failure of hormonal production in the hypothalamus, the pituitary, and the ovary. Ovarian failure is the most serious defect. In addition to the deficient hormonal production, follicles may be missing or inadequate. These defects are at present not treatable. The term 'premature menopause' is often used if ovarian failure occurs before the age of 40. Hormonal defects of the hypothalamus and the pituitary may be a manifestation of psychosomatic problems as a result of insufficient ability to cope with conflicts. This in turn may interfere with the centers bordering the hypothalamus. Problems of this kind are usually reversible. Additionally, releasing and gonadotrophic hormones can be substituted. Cyclic application of releasing hormone (Gn-RH) is simulated by small pumps or nasal spray, because in the female these hormones are released in a pulsative fashion. The easiest way to induce releasing hormone secretions is by interfering with the feedback mechanism via antiestrogens (clomiphene). When high production of prolactine blocks ovulation an inhibitory substance, called bromocryptine, is therapeutically available. This is true also for FSH and LH but these represent expensive forms of treatment.

Fallopian tube obstruction is in the majority of cases the result of tubal infection. Infection of the fallopian tube results in destruction of the tubular lining, thereby blocking the passage of the egg. Endometriosis may be another cause of infertility. In this disorder implants of endometrium are shedded through the fallopian tube into the small pelvis. These endometrial implants institute fibrous reaction, which explains infertility through interference with tubal motility by scar tissue formation. In some instances endocrine disorders may be present as well as cicatrical or inflammatory problems. Alternatively, fallopian tube obstruction can be the result of tubal sterilization.

Diagnostic, tubal patency excavations by inflating the uterus with CO_2 (Rubin test) are used less frequently now than formerly. If, instead of CO_2, a radiopaque dye is injected through the cervical canal into the uterus, the test is called a hysterosalpingography. Under X-ray scanning, the occlusion may be identified. Direct visualization techniques such as laparoscopy and the recently developed uterine endoscopy have increased considerably in popularity. For laparoscopy, a fiberglass optic tube is inserted through the umbilicus and the fallopian tubes may be observed, including the release of a blue dye from the fimbria, if it is injected through the cervix. Modern technology allows for minor surgical manipulation, for instance ovarian biopsy or freeing of adhesions.

Pathology of the uterus: In general, the uterus consists of two structures. The cervix, being the lower part, consists of connective tissue and is penetrated by the cervical canal, lined by glandular tissue. These glands produce cervical mucus, which is an important barrier against infections. The production is under the control of the sex steroids. Estrogens make the mucus watery and penetrable for sperm, progesterone decreases concentration and structure and prevents sperm migration. The condition of abnormal mucus is not well understood at present.

The upper and lower parts of the uterus consist of myometrium and the cavity, lined by the endometrium, which differs considerably from the glandular part of the cervical canal. Malformation of the uterus is not as frequently associated with infertility as one would imagine. Tumors of the uterus (fibroids) are seldom associated with infertility. This is also true of scars in the endometrium. The upper end of the cervical canal is narrow and called the internal os. Widening of this structure is termed 'incompetence' and may be a reason for inability to carry pregnancy to term.

Treatment of fallopian tube constriction may succeed by the use of one of two procedures, microsurgery and in vitro fertilization. Reconstruction of the isthmic part of the tube by microsurgery typically yields better results than operation at the uterotubal junction. Previous surgery (sterilization) is more suitable to operative correction than an operation for the results of previous infectious process.

Transplantation of a fallopian tube or the ovary has failed to be effective. This explains why in vitro fertilization has become successful, to a degree that it may be considered as an alternative method to microsurgery of the fallopian tube. In fact, this method has been described as creating an artificial fallopian tube. One ovum (or several) is extracted by laparoscopy and exposed in a test tube to spermatozoa. After fertilization has occurred in vitro one or more zygotes are implanted into the uterus. This technique has given rise to a variety of ethical questions. Most important from an ethical point of view is the fact that the egg can be from any female, the sperm from any male, and that the zygote can be implanted into any female.

Corrections of malformations of the uterus have been performed with a high rate of success. This applies also to the removal of fibroids and scarring in the endometrium (Asherman syndrome) by hysteroscopy or a too-vigorous curettage.

Repair of an incompetent cervical os is done by a variety of techniques whereby sutures are placed around the cervix (cerclage). The suture is cut at the onset of labor. Treatment of abnormal cervical mucus may be accomplished by estrogenic hormones and antibiotics, but the results are questionable at best.

Functional Defects in the Male

Spermatogenesis depends on hypothalamic or pituitary hormones. Daily, new spermatozoa are formed. However, the germinal epithelium is much more easily destroyed than the corresponding cells in the ovary. Infection and chemotherapeutic agents are cases in point.

Germinal epithelium may be absent in some congenital syndromes (Klinefelter syndrome), or acquired when the testes are retained in the abdominal cavity (cryptorchismus). Although spermatogenesis depends on hormonal action, there is no feedback mechanism involving germinal epithelium. Aspermia results from destruction of the germinal epithelium, congenital absence or destruction of the ejaculatory ducts by infection (gonorrhea).

Abnormal sperm may be the result of infection, chemotherapy, irradiation, trauma and/or tumors. Varicosis of the veins draining the scrotum is considered to be a factor (varicosities) as well as tight trousers or jeans.

Surgical correction of vas obstruction is less satisfactory than tubal reanastomosis. A low sperm count (oligospermia) can result from insufficient hormonal stimulation or destruction of the germinal epithelium. The normal sperm count is in the range of 40 million per ml. Pregnancies have occurred with sperm count as low as 10 million, though such cases are rare.

Interference with the nerve supply of the accessory glands may be the result of para-aortic lymphnode dissection, or the intake of drugs for the control of hypertension. This explains erection and subsequent orgasm without ejaculation. On the other hand, erection and orgasm may be impaired, yet ejaculation accomplished by masturbation or electric stimulation in such patients as those with multiple sclerosis or spinal cord injuries. Under these conditions sperm may be ejaculated into the bladder (retrograde ejaculation). Sperm agglutination and aggregation, as the result of the presence of sperm antibodies, are less frequent than previously assumed.

Treatment of male conditions is less successful than that of female conditions. Correction of varicose veins (surgical ligation) results in pregnancy in more than 50% of cases. Clomiphene and gonadotrophic hormones have been used with variable success. In the case of low sperm count, but with normal motility and morphology, insemination of the female with a concentrate of the male partner's semen (homologous insemination) has been highly successful. This is also true for sperm obtained from the structure of accessory glands. Inability to achieve an erection is more frequently related to psychological factors than to organic causes. Noticeable is the effect of alcohol and a variety of drugs. Treatment of spermatozoal antibodies is difficult. Use of condoms as a contraceptive for six months has been advocated, but success rates are unclear.

ETHICAL CONSIDERATIONS

Contraceptive techniques that interfere with implantation are not considered abortifacient by most societies. Some decades ago, many national societies and colleges of obstetrics and gynecology declared that pregnancy starts with the culmination of the implantation process. This was taken into law in many Western countries. It provided a means of differentiating between a contraceptive agent and an (early) abortifacient. Both the IUD and the pill do interfere with the implantation process but by definition this was of no consequence as far as abortion inducement is concerned. However, this definition has been shown of great consequence for in vitro fertilization (see below).

In regard to contraception, interference with sperm migration may be considered equivalent to ovulation inhibition. An agent having these characteristics is accepted by most (excepting Roman Catholic clergy) as a true contraceptive agent. It may be remembered that until implantation either reduplication of one organism or recombination of two into one may occur. Hellegers has brought one very interesting thought into the discussion: "If during two fertilizations two souls have originated, and if one body contains only one soul, then we see cases now in which the soul must have disappeared without the death of a fertilized egg" [7].

The 'morning after pill' cannot be considered abortifacient because the egg does not reach the endometrial cavity for up to 7 days after ovulation. Only when the endometrium is shed with an egg in the stage of implantation could the agent be termed abortifacient. Presumably it is necessary to differentiate between early and late contraceptives. It seems to make a difference whether an agent destroys the blastocyst (by chemical or physical means) rather than blocking sperm migration or inhibiting ovulation.

Other ethical considerations in regard to contraception are concerned with the consent of minors. Throughout the ages, debates have been held on what the appropriate age for self-determination for minors should be. This applies especially to sterilization procedures, since this technique is considered irreversible. Many physicians refuse to sterilize a 20-year-old single woman on demand since they feel that even a human being of this age cannot foresee the consequences such an action may have for her life 10 or 20 years later. This seems to be supported by the fact that the demand for refertilization is sharply increasing. Even more difficult is the decision for sterilization of a mentally retarded adult or minor.

The ethical dilemma in regard to abortion can only be treated briefly while remaining within the limits of this chapter. There seems little doubt, especialy in regard to in vitro fertilization, that human life starts with fertilization. If one argues about individualization after implantation has occurred, human life is a continuous growing. Spontaneous heart beat is present in the sixth week.

There is no cesura during pregnancy that makes a time or a given stage of pregnancy more prone to interruption than any other time of fetal development. It may be questioned whether birth is even such a cutting point. Hellegers has raised the question whether the placenta is a fetal organ of spontaneous respiration or a respirator. Without an answer to this question, it follows that there is no biologic distinction about sustained human life and it is up to society to define, more or less arbitrarily, when it is willing to protect human life in or ex utero. To require other qualities besides its human presence (such as the externally conferred privileges of 'having value', 'having dignity', 'having a soul', or being 'a person') steers the discussion into an emotional outbreak. This is exacerbated when functions related to humanity, like feeling pain, thoughts about giving and receiving love, and so on, are introduced. If one asserts 'quality of life' or 'meaningful life' then, of course, the question arises as to who is responsible for the decision making.

There is little disagreement that abortion is justified on grounds of serious conflicts between a maternal and fetal interest. This is taken to be a medical indication for abortion. In many countries the law contains the term 'criminal indication', which means termination of pregnancy after rape. When a legal term like 'social indication' is used, it points to the insecurity of a given society. Indeed, every termination of pregnancy for a fetal indication, or so-called 'eugenic indication', is a social indication because nobody knows whether a genetic aberration (e.g., mongoloidism) results in a happy or unhappy life (if anybody knows what this means when applied to persons of limited speech ability). This is, of course, not the reason for the termination but the burden that is placed on parents and society.

This has special bearing on the so-called prenatal screening or prenatal diagnosis. This term was used originally to describe a technique whereby amniotic fluid is obtained (amniocentesis) and fetal cells cultured. Genetic aberrations could then be diagnosed. In the meantime, ultrasonography has reached such perfection that a great variety of minor anatomical malformations are detectable. From this a problem arises: a physician can be held liable for not advising the patient of the availability of such techniques. One of the first instances of this was the Curlender case where a defective infant recovered damages for being born to a 'wrongful life'. The number of chromosome 21 trisomic children increases after the maternal age of 35 years from 1% to 3–5% by the age of 42. Therefore many laws state that the mother has a right to prenatal diagnosis. The 'fundamental right to be born as a whole functional being' is protected by law. What, however, happens if there is a misdiagnosis and the fetus is in fact normal? Can the physician then be sued? This makes one wonder why one should go to all the trouble and expense of doing intrauterine tests that might harm the fetus for inspection of the prospective infant when diagnostic tests at birth would be much easier. An infant with serious birth defects could be terminated at that time (cf. [3]), as was the practice in ancient Greece. Proponents of this viewpoint suggest that a newborn baby should not be considered legally human until certain standards of 'normality' have been assured. They even consider this procedure as more humane than prenatal diagnosis because children with minor defects who would otherwise be terminated would be allowed to live. But who terminates the 'unnormal' after birth and who sets the standards? Does an unwanted girl rather than a wanted boy already belong to an 'unnormal' category as some East Indians believe? This is the problem of sex selectivity, which may be more disastrous to future life than overpopulation.

There is considerable disagreement regarding fetal experimentation, and this overlaps with questions regarding the use of fetal organs for transplantation.

In vitro fertilization (IVF) is associated with a variety of ethical concerns. Most Western societies have not legally solved the problems associated with donor insemination: the acceptability of sperm banks, selection of sperm, and indications for use. Ramsey objects to the use of IVF on the basis that infertility is not a disease and therefore the method is not to be considered a treatment *per se*. This, however, must then also apply to tubal surgery. In both instances one may put others at risk for one's own ultimate good. This is considered an extremely conservative position held only by the Catholic Church, which attaches to the act of love such a significance that the technical problem of masturbation for semen procurement blocks acceptance. The Geneva Convention states that it is the right of every woman to have the opportunity to have children of her own, which seems to be one of the exaggerations of the WHO definition. If the woman has the right, what about the man? Is he solely the sperm donor?

The method of IVF seems to be acceptable to the majority of physicians if the oocyte is obtained from the wife, the sperm originates from the husband, and the fertilized blastocyte is retransferred to the wife again. Once the embryo is implanted the development is natural, ending in the delivery of a fetus having the genetic structure of the two parents, just as would result from a 'normal' fertilization process.

The low success rate of IVF can be compared with the large failure rate of nature. It is expected that only 30 oocytes out of 100 exposed to fertilization lead to a living child, regardless of the fact that 50% are implanted. In most instances two or more oocytes are expected from a hormonally stimulated cycle. The wastage problem is solved if only two oocytes are transferred and any additional cells undergo natural ovulation and degeneration. However, this is exactly the point at which misuse is programmed. Some feel that additional oocytes should be placed in an oocyte bank and used for transfers to subsequent cycles. The questionable success of commercial sperm banks should engender caution.

The supporters of IVF already accept what others call a misuse: the acceptance of donor sperm, or a donor blastocyst, or the involvement of a surrogate mother. These procedures result in a half-genetic child and are considered identical to heterologous donor insemination. The question whether a woman can guarantee her product (the quality of her gamete) is the least important problem in question. (Is producing the egg equivalent to masturbation or not, and if not, why not?) If the sperm used is from a male donor and the egg is from a female donor, this is considered by proponents of IVF as equivalent to adoption. For the opponents of IVF, this method is claimed to obliterate any family structure. Since the technique is already available to produce twins, one can ask why one should assume the burden of having two pregnancies instead of one. The supporter of IVF will maintain that selection will improve the human race, a questionable goal in view of the lack of standards for human values.

Legal questions involving IVF are in no way resolved. This applies especially in the case of surrogate mothers. What is her relationship to the genetic father or the genetic mother – who is, indeed, biologically speaking, the *mother*? What happens if during pregnancy either the surrogate mother, or the genetic parents as employer, change their minds? Is abortion then appropriate? These are only some of the unsolved questions. One feels that in the case of an ovum donor/surrogate mother and the genetic mother, the two should not know each other. However, if one recalls the law in Scotland where an offspring has the right to clarify its own origin, then who does this apply to, the genetic or the surrogate mother?

Even more perplexing is the combination of IVF and genetic engineering. The attempt to improve the human race is to play God, but it also can be used to play Satan. The requirements of Freedman and Rublin are still the basis for decisions regarding such procedures: for acceptable genetic treatment of a human defect, it is required that gene therapy replace the functions of the defective gene segment without causing deleterious side effects either in the treated individual or in his future offspring. Although at present every attempt is made to preserve the normal physiologic state of the oocyte or sperm in order to improve implantation results, this may not be the case as soon as the technique becomes standardized. Since pregnancy begins with implantation, the fertilized blastocyst is not protected by any law and therefore manipulation at this stage is uncontrollable.

It may be of help to have ethics committees on a local or national level. However, legal tradition seems to be very difficult to establish and will be different from country to country. Who is controlling the controllers? – a Pandora's box of unsolved questions. A group of investigators involved with IVF may be afraid that freedom of research will be limited by laymen sitting on ethics committees. However, there is an equally eloquent group that feels that technology has gone too far and needs now to be controlled. The present situation is to allow the methodology to exist, and to prohibit as exceptions cases of misuse. Might it not rather be preferable to ban the methodology and allow for a few acceptable exceptions?

Der Universitäts–Frauenklinik,
Albert Schweitzerstr. 33,
Münster, West Germany

BIBLIOGRAPHY

1. Beller, F. K. and Quakernack, K.: 1980, 'Fragen zur Bioethik', *Geburtshilfe und Frauenheilkunde*, **40**, 142–145.
2. Beller, F. K., Schweppe, K. W., and Wagner, H.: 1984, *Intrauterine Kontrazeption*, Edition Medizin, Weinheim, Deerfield Beach, FL, Basel.
3. Callahan, D.: 1980, 'Shattuck Lecture – Contemporary Biomedical Ethics', *New England Journal of Medicine* **302**, 1228–1234.
4. Edwards, R. G.: 1974, 'Fertilization of Human Eggs in Vitro: Morals, Ethics and the Law', *Quarterly Review of Biology* **49**, 3–8.
5. Edwards, R. G. and Sharpe, D. J.: 1981, 'Social Values and Research in Human Embryology', *Nature* **231**, 87–91.
6. Hanker, J. P., Nieschlag, F., and Schneider, H.: 1981, 'Frequency-modulated Pulsatile LH-RH Substitution in Hypothalamo-amenorrhic Women', *Acta Endocrinologica* **96**, Suppl. 240, 75–76.
7. Hellegers, A. E.: 1978, 'Fetal Development', in T. Beauchamp and L. Walters (eds.), *Contemporary Issues in Bioethics*, Dickenson Publishing Co., Encino, pp. 194–198.
8. Johnston, J. *et al.*: 1981, 'In Vitro Fertilization: The Challenge of the Eighties', *Fertility and Sterility* **36**, 699–706.
9. Nieschlag, E. *et al.*: 1982, 'Reproductive Functions in Young Fathers and Grandfathers', *Journal of Clinical Endocrinology and Metabolism* **55**, 676–681.
10. Pauerstein, C. J. and Bartke, A.: 1982, 'Embryology, Maturity and Anatomy of

the Reproductive System', in R. Shain and C. Pauerstein (eds.), *Fertility Control*, Harper & Row, Hagerstown, pp. 3–20.

11. Stone, O.: 1973, 'English Law in Relation to AID and Embryo Transfer', in G. Nolstenhome (ed.), *Law and Ethics of AID and Embryo Transfer*, Associated Scientific Publishers, Amsterdam, pp. 69–76.
12. Tatum, H.: 1972, 'Intrauterine Conception', *American Journal of Obstetrics and Gynecology* **112**, 1000–1023.
13. Tietze, C.: 1970, 'The Condom', in M. Calderone (ed.), *Manual of Family Planning and Contraceptive Practice*, 2nd ed., Williams and Wilkins, Baltimore, pp. 232–244.

SECTION II

SEXUALITY AND SEXUAL CONCEPTS

ALAN SOBLE

PHILOSOPHY, MEDICINE, AND HEALTHY SEXUALITY

INTRODUCTORY REMARKS

'Healthy sexuality' is a chameleon concept. On the one hand, it seems that nothing could be easier to understand, or that there is nothing to be understood. The concept of 'healthy sexuality' does not arise naturally in our biographies. We learn at our mothers' knees (or nearby) what is right, decent, proper sexual behavior. She uses the language of morals, not the language of health. So there is nothing to be understood. When health is mentioned at all, we are told to keep our body parts clean and we learn about venereal disease. So 'healthy sexuality' is easily understood in terms of soap and penicillin.

On the other hand, 'healthy sexuality' is part of the conceptual stock-in-trade of the professional and the academic. It arises naturally in the helping professions, among social workers, educators, theologians, psychologists, and physicians. Because the central concept of the academic discipline of bioethics is 'health,' the concept 'healthy sexuality' is also a natural amalgam to be explored by philosophers (like 'genetic health'; see [36]). The distinctive philosophical issues are ontological and epistemological. Is healthy sexuality a set of behaviors, and if so, how are we to decide which sexual behaviors are the healthy ones? Or is healthy sexuality marked by a type of mental state that can accompany, as a cause (intentions, attitudes) or as an effect (pleasure, satisfaction), any behavior at all? The helping professions are not immune from these difficulties; implicitly or explicitly they work with models of healthy sexuality that presuppose answers to these questions.

The academic rigorously scrutinizes what we learn at our mothers' knees, sometimes beginning afresh. More often, the moral retains some influence on this intellectualizing. Either the moral is repudiated altogether as irrelevant and regressive; often it is repudiated with such vehemence that we gasp at the savagery of the protest. Or we find that, at the logical bottom, our mothers' teachings about what is right and decent are repeated, albeit secularized, cast into enlightened terms. We are so used to thinking about sexuality in moral terms, almost as if

E. E. Shelp (ed.), Sexuality and Medicine, Vol. I, 111–138.

questions of sexual behavior exhausted the realm of moral discourse, that our accounts of 'healthy sexuality' do not break free.

Our mothers, however, have recently been instructing us under the gaze of the helping professionals. Some of what we learn at her knee has become infected with the thought and language of the intellectual classes (see [25]). We now hear from her, as well as from the helping professionals, a slew of pronouncements. We hear plenty of reasons *not* to engage in sex, both moral and health reasons. When sexual diseases (herpes, AIDS) are uncontrollable, the moral reasons are buttressed by an appeal to an avenging nature. When advances in medicine make the appeal to physical health ineffective (e.g., post-syphilis), deeper psychological or metaphysical health considerations are adduced. The humanist movement of the last hundred years has replaced the moral critique of sexuality with a no-less relentless medical critique. Yet we also hear from other helping professionals reasons *to* engage in sex of all kinds, reasons based not so much on positive moral considerations but on the bankruptcy of the old moralities, and reasons that extol sexuality itself as a standard of the healthy. Judging sexual behavior by a criterion of 'healthy sexuality' is turning passé, in favor of a definition of health in terms of sex: having it is healthy, not having it is not. 'Healthy sexuality' is becoming redundant.

THE SCOPE OF MEDICINE

What *can* medicine say about 'healthy sexuality'? If we understand medicine to be the science of the functioning, pathology, and therapy of the body and its parts, we have one position: (1) healthy and unhealthy sexuality are divided organically. Richard Lumiere's book [28] tells us almost the full capacity of medicine: *Healthy Sexuality and Keeping It That Way. A Complete Guide to Sexual Infections*. 'Healthy,' here, is 'hygienic.' Healthy sexuality does not transmit venereal disease. We learn that heterosexuals should not engage in vaginal intercourse right after anal intercourse without washing the penis. In addition, healthy sexuality includes the absence of trauma, tumors, neuropathology, and toxic or metabolic disorders, conditions that might disrupt the effective working of the sexual parts (e.g., some cases of insufficient lubrication).

Students of Szasz[1] will say that there is nothing more to sexual health; it is only a narrowly medical-biological concept. But others will balk, and advance two other positions. (2) There is much more that medicine can do in illuminating 'healthy sexuality'; infections and the like do not exhaust the medical approach. Psychiatry is a legitimate branch of medicine and has its own contribution to make. After all, the science of the functioning and pathology of the body includes the science of the brain. Of course, 'functioning' is now extended to thoughts, motives, desires, and other mental states and phenomena. But in some sense we are still concerned with the body, in particular the integration of the brain, behavior, and the sexual organs.[2] (3) Medicine can take us only as far as infections, tumors, and the like, and these items are part of what we are concerned with in understanding healthy sexuality. But there is much more to healthy sexuality; this extra, however, is outside the domain of medicine. Psychologists, social workers, religious advisers, even moral philosophers, have their own unique contributions to make. If psychiatry wants to contribute, it can do so, but not exactly as a branch of medical science.

Position (2), more than position (1), preserves the centrality of medicine by retaining the significance of psychiatry, and we should expect it to be defended vigorously by the medical establishment. But medicine runs the risk of committing a gross *non sequitur*. Medicine can seduce itself into arguing that even though it is empowered to deal *only* with the body (as in position (1)), it is empowered to deal with *anything* concerning the body, which is the whole domain of human life. Position (2) is also liable to this reasoning, for in expanding the concept of the 'functioning' of the body to include the brain and therefore any brain-associated phenomena, again the whole domain of life is covered. There is no doubt that medicine has often committed this mistake.[3]

The World Health Organization's definition of health as "a state of complete physical, mental and social well-being and not merely the absence of disease or infirmity" ([3], p. 83) is often accused of making or warranting this *non sequitur*. Robert Veatch, for example, complains that the definition

> means that . . . the medical profession is the one to turn to for technically competent help in such failures in well-being as marriage problems, poverty, and unanswered prayers ([48], p. 73).

But it is unfair to accuse the WHO declaration of making the *non-sequitur*, because WHO also says that "the extension to all peoples of the benefits of medical, psychological, and related knowledge is essential to the fullest attainment of health" ([3], p. 83). Thus, WHO proposed a broad notion of health having a variety of ingredients, and it is clear that medicine is restricted to some of these while psychology and 'related' fields cover the others. It is easy enough to match the WHO definition of health with this additional claim, and conclude that WHO meant that medicine covers physical, psychology covers mental, and 'related knowledge' (sociology?) covers social well-being. WHO might be guilty of proposing a too-inclusive notion of health, but it is not guilty of proposing a too-inclusive role for medicine.

It is understandable, however, that some take the WHO definition of health as justifying a predominant role for medicine in every aspect of life. 'Health' is foremost a medical concept, and the broad notion of health proposed by WHO does not involve a mere metaphorical use of the term. It could be concluded, then, that WHO thinks that all human problems are health and *therefore* medical problems. As a result, one might worry that the WHO definition would have, as Daniel Callahan claims, "the practical effect of blurring the lines of appropriate authority and responsibility" ([11], p. 78, p. 82). But if we take into account WHO's additional remark, the force of its using 'health' as 'well-being' is diluted. Engelhardt can define 'health,' if he wants, as the absence of any condition that causes people to lose the "ability to act freely and rationally" ([14], p. 266; [15], p. 216), and thereby defend the WHO definition ([14], p. 267), without fearing that he is thereby suggesting that physicians automatically become philosopher-kings. The definition of health by itself does not entail what discipline takes charge. At the same time, it is misleading for Callahan to speak about the 'appropriate' boundaries on disciplines, as if these boundaries are to be established by philosophical reflection. The boundaries are set, rather, by the competition among the disciplines themselves (see below, pp. 127–129).

PHILOSOPHICAL PROBLEMS IN MEDICAL JUDGMENTS

> . . . health and disease in a body are admittedly different and distinct. . . . Desire is one thing in a healthy body, and it is another in one diseased . . . It is fine and even proper to yield to the good and wholesome wants of every body, . . . but it is bad to yield to the noxious and evil desires. . . . For this is what medicine is, . . . a knowledge of the forces of Love in the body. . . . A person who can accurately diagnose whether the noble or the vulgar love is functioning . . . and can interchange them is a master therapist.
>
> Eryximachus
> (Plato, *Symposium*, 186 b-d)

It is fruitful to examine what medicine *has* said about 'healthy sexuality.' Has medicine improved on Eryximachus' performance? An editorial in the *British Medical Journal* [1] tells us:

A doctor's special responsibility is to distinguish the healthy from the unhealthy and to teach the facts. And though he must be understandably sensitive about interfering in moral problems he should not shrink from giving guidance on the medical and biological components of them where people's health is concerned. . . .

This sounds reasonable. Yet the subject of the editorial is *pornography*. What can a physician, who is concerned with 'the facts' and only the biological components of moral behavior, say about pornography? That masturbating with pornography for five hours can cause blisters, or that one might catch a disease from the seat-cushion in a movie theater? No. The *Journal* claims that

the flourishing trade in pornography serves to distort the loving and biological expression of the sexual instinct. In so far as it succeeds in doing that it impairs the health and well-being of the people in its thrall.

The sleight-of-hand is inexcusable. It is one thing to claim that pornography distorts the *loving* expression of sexuality. Perhaps those who consume pornography come to separate sexual experience and love (or have already done so), but how this can be interesting from a medical perspective is unclear.[4] It is quite another thing to claim that pornography distorts the *biological* expression of sexuality. If so, it is reasonable for the *Journal* to claim that it impairs health. The *Journal*, however, never tells us what goes wrong biologically when one is in the thrall of pornography. That a penis might erect or a vagina lubricate in response to visual or linguistic stimuli does not seem to be biologically anomalous. Nor is there evidence that pornography destroys the ability of a penis to erect or a vagina to lubricate in the presence of another person. Is the *Journal* assuming that the expression of love is the biological function of sexuality? At least that bold thesis makes the *Journal*'s position coherent. We shall return to the question of the function of sexuality.

The physician Max Levin provides another example; he writes:

oral-genital and other sex acts are healthy and legitimate, provided that both parties enjoy them, and provided they are practiced out of love and devotion . . . and . . . provided they are no more than a part of the foreplay that builds up erotic tension to climax in coitus ([26], p. 624).

Levin gives us three conditions, no less, for 'healthy' *and* 'legitimate' sexuality. Does Levin want to say that the healthy and the legitimate are the same, or that whatever makes sex healthy also makes it legitimate? If we focus on 'healthy,' none of Levin's three conditions are perfectly happy ones. For example, on his view fellatio that occurs in a context of devotion is healthy, but otherwise unhealthy. That implication is much too counter-intuitive to be acceptable without a good deal of sophisticated argument. Because these conditions fail as criteria of healthy sexuality, it is plausible to think that what Levin has done is to provide conditions that can only be understood as moral criteria, and disguised (not very effectively) as medical. Nevertheless, one wonders whether some logical requirement of analyzing 'healthy sexuality' leads one to proposing conditions of the same sort.

Another example comes, as it were, right from the horses' mouths: four short essays on distinguishing 'healthy' from 'sick' sexuality, which appeared in *Medical Aspects of Human Sexuality* [31]. Judd Marmor begins with no reservations about the difficulty of the task, yet easily proclaims that "Healthy sexuality tends to be discriminating as to partner choice; neurotic patterns tend to be nondiscriminating" (p. 67). One value judgment involved here is the judgment that partner-discrimination is the mark of the healthy; a second value judgment is involved in deciding how much promiscuity is required for nondiscriminatory partner-choice.[5] Robert Gould cautions us that the distinction between healthy and sick sexuality "is not only subjective but philosophical as well" ([31], p. 75). It would seem, then, that Marmor, Gould *et al.*, have no business opining professionally on the matter; this would seem to be a case of the 'generalization of expertise' [47]. (It never occurred to me that tacking on 'philosophical' to 'subjective' was akin to adding insult to injury.) Undaunted, Gould proceeds to claim that sadomasochism is 'sick' because it is 'dehumanizing and degrading.' For Richard Friedman and Robert Spitzer, "nosological schemes involve value judgments" ([31], p. 75); according to *their* values the dehumanization of sadism makes it a 'disorder' (p. 76). Finally, Thomas Clark, after disabusing us of the "idea that any sexual behavior *per se* is either sick or healthy" ([31], p. 76), ignores his sound advice: "sexual behaviors which violate, exploit, or damage another human being cannot be viewed as sexually healthy" (p. 77). These physicians have repeated Levin's mistake, confusing judgments of 'healthy' and 'legitimate.' What is amazing is the fact that our writers insist on the subjectivity of distinguishing between healthy and unhealthy sexuality, yet agree that sadomasochism is 'sick.' Is this agreement evidence that cultural bias against these sexual practices infects the professional physician, or that judgments of healthy sexuality are not as subjective as Marmor *et al.* would have us believe?

Finally, when thinking about sexual health, one must turn to Masters and Johnson, especially if one dislikes psychodynamics. But reading *Human Sexual Inadequacy* [32] is painful for lovers of conceptual clarity. In the Preface and elsewhere, 'dysfunction' and 'inadequacy' are used interchangeably; neither appears in the Index. In Chapter One, 'Therapy Concepts' (pp. 1–23), the place where we expect to find it, there is no definition of 'inadequacy' or 'sexual dysfunction,' the central concepts. What the authors do, instead, is to define *each* sexual dysfunction, apparently satisfied with an ostensive definition. Masters and Johnson pay the price. For example, during their discussion of heterosexual male impotence, they write about the

> [n]egation of the young male's potential for effective sexual functioning . . . ([32], p. 137).

But the phrase 'effective sexual functioning' is redundant: all sexual dysfunctions, for Masters and Johnson, involve some sort of ineffectiveness in carrying out sexual activity.

Nevertheless, talk of sexual dysfunction in medicine has caught on. Irving Bieber, who complains that ". . . the term 'sexual deviation' is ambiguous, vague, and not useful as a diagnosis or as a nosological category" ([45], p. 1210), proposes to replace it altogether with the terminology of dysfunctions. In a *tour de force* of inarticulate medical linguistics, Bieber says:

> Masters and Johnson used criteria that qualified as functional and dysfunctional to classify sexual disorders, and they introduced the term "sexual inadequacy." Under this rubric, they included frigidity and sexual impotence. . . . I suggest that homosexuality be characterized as a type of sexual inadequacy since most homosexuals . . . cannot function heterosexually ([45], p. 1210).

'Sexual dysfunction,' however, is no less ambiguous and vague than 'sexual deviance'; note how Bieber glides so smoothly into calling homosexuals 'inadequate.' 'Dysfunction' *sounds* more medical than 'deviance,' an advantage for physicians who want to include sexuality in their domain.

It is instructive to consider two criticisms of Bieber. Richard Green agrees with Bieber that 'sexual dysfunction' should be the medical concept of choice, but disagrees that homosexuality *per se* is a sexual dysfunction. On Green's view, 'sexual dysfunction'

> would include the heterosexual or the homosexual who finds it difficult to maintain desired object relationships, who compulsively uses sexuality to ward off anxiety or depression, or whose sexuality typically leads to depression or anxiety. . . . With this new classification, psychiatry would have an objective basis for categorizing sexuality that is free of cultural bias . . . ([45], p. 1214).

Clearly, Green is going well beyond Masters and Johnson's *non*psychodynamic use of 'sexual dysfunction,' for he includes not only 'physiological dysfunction of psychogenic origin' but also 'ego-dystonic,' 'psychogenic distress' as indicative of sexual dysfunction. One wonders, then, whether Green is merely appropriating for psychiatry a term that carries some clout. Against Bieber, Green maintains that even though the homosexual cannot 'function' heterosexually, he or she can still 'function' homosexually, and as long as such behavior is not 'ego-dystonic' it is not a sexual dysfunction. But Green's pronouncement that his criterion is bias-free is questionable. If he means that the inability to maintain relationships is a dysfunction *only when* one desires to maintain relationships, he will get both complaints and praise for this judgment. For Green, dysfunctional sexuality is that which is accompanied by, or is intended to ward off, depression or anxiety; to be consistent, shouldn't he insert the proviso here as well, so that the use of sexuality to ward off depression is dysfunctional only when one would rather not use sexuality that way? We hesitate when Brenda Wiewel says that "true sexual liberation will occur when a woman feels completely comfortable with her sexual response no matter what it is" [49]. Is 'feeling comfortable' really the *whole* story?

C. W. Socarides protests that Bieber's replacement of 'sexual deviance' with 'sexual dysfunction' is a disaster:

> Scientific knowledge is . . . damaged when attempts are made to classify homosexuality simply as "sexual dysfunction," a term regularly applied to loss of erection, premature ejaculation, retarded ejaculation, or total impotence. These impairments constitute disturbances of the standard male-female pattern. . . . Individuals unable to achieve sexual release within the standard pattern . . . turn to modified patterns for orgastic release, and these constitute sexual deviations. Thus the immutable distinctions between sexual deviations and sexual dysfunctions cannot be semantically blurred without incurring formidable scientific chaos ([45], p. 1213).

Sound the alarm! – Bieber plans to change the 'immutable.' Yet the passage does highlight a tangle in the talk of sexual functioning. In one sense, both the heterosexual and the homosexual can function sexually, i.e., engage in sexual activity effectively. In another sense, the homosexual is dysfunctional because he or she cannot function heterosexually, i.e., cannot engage in heterosexual activity effectively.[6] But we ask: why is a heterosexual who cannot function homosexually not dysfunctional as well? Bieber, in classifying homosexuality as a dysfunction, assumes that heterosexuality is superior to homosexuality. Socarides is merely making this judgment explicit, and is insisting, in virtue of this judgment, that there is a difference in kind, not merely in degree, between inadequately consummated heterosexual relations (premature ejaculation) and never-desired heterosexual relations. Of course, Socarides shuffles around when he calls heterosexuality the 'standard' pattern, as if with that word he can avoid the embarrassing question of the status of his judgment.

One attempt to define 'sexual dysfunction' brings out an important dimension of this concept:

> Two types of sexual problems are examined. The first, which we have labeled "dysfunctions," includes erectile and ejaculatory problems in the husbands and arousal and orgasmic problems in the wives. The second type of problem we have labeled "difficulties" and includes the following complaints: "partner chooses inconvenient time"; "inability to relax"; "attraction to person(s) other than mate"; . . . The distinction between "dysfunctions" and "difficulties" is to some extent an arbitrary one that we have chosen *because* we believe it preserves the distinction between problems of "performance" . . . and problems that are more related to the emotional tone of sexual relations ([19], p. 112; italics added).

The distinction between 'dysfunction' and 'difficulty' is slippery. Suppose that a man's inability to maintain an erection with his wife (a dysfunction) is due to his being bored after six years of marriage (a difficulty). This is not a medical problem; there is nothing biologically wrong with a male that will not respond to another repetition of the same stimulus. Or suppose that a woman's inability to have an orgasm is due to her husband's being a dolt of a lover. Problems these are, but they don't seem to be exactly health, let alone medical, problems. Or so it would seem. If medicine can capture this territory, these problems *are* medical problems (see pp. 127–129). The definition of 'dysfunction' as a condition affecting sexual performance is revealing. The inclusion of sexual performance problems into the domain of medicine makes sense in the context of 20th-century industrial society. That sexual dysfunction is conceived in terms of performance, and that the sexual problems are erectile-arousal, are understandable concomitants of the emphasis in business and economics on productivity. The firm, the machine, the worker who do not perform are not healthy firms, machines, workers. The man who cannot penetrate, the woman who cannot have an orgasm,[7] are machines in disrepair. In a society that is obsessed with sex, sexual performance problems stand out as worthy of treatment by those who trade in the repair of the body.

The concept of 'sexual dysfunction,' it can be argued, shows us that the analysis of 'healthy sexuality' requires evaluations. Consider this straightforward explanation of 'sexual dysfunction':

> The penis is rubbed and it doesn't get hard. . . . A couple copulate for a long period of time and the husband and wife don't reach a climax. Obviously, something has gone wrong, so that the expected response is not produced by a particular stimuli. When this happens, the condition is called a *sexual dysfunction*, which really means a malfunctioning of the human sexual response system. It means that a person hasn't reacted as one would normally expect. . . . [T]he sex organs do not respond as they should ([39], pp. 22–23).

We 'normally expect' that a rubbed penis will erect; a rubbed penis that does not erect is not responding 'as it should'. The point seems to be that 'healthy sexuality' is to be understood in terms of the smooth operation of the organ's mechanical process. Yet the words 'normally' and 'should' suggest that it is necessary to appeal to evaluations in analyzing 'sexual dysfunction' and therefore 'sexual health.' We shall see that this claim has been debated in the professional literature.

WHY "HEALTHY" AND "HEALTHY SEXUALITY" ARE EVALUATIVE

Joel Feinberg argues that health judgments about the body and its organ-parts are not entirely factual, but are in part evaluative. Apparently, the

> ascription of functions to component parts . . . is a wholly factual matter consisting of . . . a description of the part's effects and . . . a causal judgment that these effects are necessary conditions for the occurrence of some more comprehensive effects ([17], p. 253).

If so, then the judgment that a component organ is diseased, the judgment that the organ is dysfunctional, will be wholly factual. "But," Feinberg continues, "the illusion of value-neutrality vanishes when we come to ascribe a function to the organic system itself" (p. 253). The point is that the 'more comprehensive effect' to which the organ contributes is the working of the body, and evaluations come into play in deciding what bodies *should* be able to do (see also Engelhardt, [15], p. 215). Thus for Feinberg:

> [i]t follows that even statements ascribing functions to component organs will not be entirely value-neutral, for the macroscopic functions for which their effects are necessary conditions will contain value specifications in *their* descriptions ([17], p. 254).

A diseased heart, for example, is a heart that interferes with or does not permit the 'proper' working of the body. Because what counts as the proper working of the body depends on "our resources, technical capacities, and purposes" (p. 255), and because claims about purposes are evaluative, health judgments about component organs are, ultimately, evaluative.

Feinberg's view makes sense, but I have three qualifications to add. First, Feinberg's conclusion is that any statement that an organ is diseased is evaluative, and not the quite different conclusion that the concept of 'health' is evaluative. That is, one can analyze 'healthy body' as "a body that is working well" and not be committed to any value judgments. One does not make a value judgment simply by asserting this analysis of the concept rather than the opposite (and wrong) analysis, 'healthy body' means "a body that is not working well." Value judgments are involved specifically at the stage when one decides what a properly working body is. But this role for evaluations is ubiquitous; it occurs whenever, having analyzed a concept, we decide whether a particular object instantiates that concept (whether it is a 'proper' *X*). Some instantiation decisions are more difficult than others ('health' vs. 'chair'), but they are distinguished in terms of their difficulty (which derives from the practical consequences of instantiation), not in terms of being evaluative.

Second, that evaluations are presupposed by 'properly working body' does not entail that judgments of the health of component organs are evaluative. *Given* a statement, even an evaluative statement, about the proper working of the body, the subsequent ascription of a function to an organ is purely descriptive. That these health judgments are *relative* to background conditions (resources, technical capacities, and purposes) does not entail that they are themselves evaluative. Again, this feature of health judgments is ubiquitous. It is by now a common view of science, that given a set of background assumptions, some of which are evaluative, subsequent hypotheses and observation sentences are empirical relative to that set. The search for empirical assertions that are independent of such a background has proven futile.

Third, it is just not the case that *all* statements about the proper working of the body are evaluative. Surely a body that is working properly is a body that is alive and kicking, rather than one the life of which is precarious or degenerating. A body that is alive and stable is a minimally healthy body, regardless of whether that body happens to be bowling or watching television (see [6], p. 60; compare Engelhardt, [15], p. 215). An organ that contributes to the maintained life of a body is at least a minimally healthy organ. Describing such organs and bodies as healthy does not require any value judgment, not even the value judgment that it is better for a body to be alive than dead. It is simply to say that a healthy body, with healthy parts, is one that has a sustained existence. Thus, there is some reason to claim that medicine, as the science of how to keep bodies alive, is the discipline in charge of making this sort of purely descriptive health judgment. Of course, it is doubtful that we can get much mileage from this narrowly circumscribed descriptive content of 'health' in understanding 'healthy sexuality.'

It is the central thesis of Christopher Boorse's analysis of 'health' that disease judgments are nonevaluative. For Boorse, "the root idea . . . is that the normal is the natural" ([6], p. 57; [8], p. 554). Healthy or normal functioning, then, is just natural functioning. To function in accordance with nature, in turn, is to function in accordance with the evolutionary natural design of an organ. Health is the absence of disease, a condition that interferes with an organ making its "species-typical contribution to survival and reproduction" ([7], p. 62; [6], p. 57; [8], p. 555). It follows that "disease judgments are value-neutral" ([8], p. 542), and that the recognition of diseases "is a matter of natural science" ([8], p. 543; [6], p. 59; [7], p. 63). In order to make health judgments all we need to do is to ascertain the species design, and we can do this for humans just as we do it for frogs.

Boorse's account of health and disease has come under some fire [9, 14, 30, 35]. These criticisms raise doubts that Boorse has seen how deeply, and in what ways, judgments of health are value-laden. Joseph Margolis writes that

> . . . since medicine in general must subserve . . . the . . . ulterior goals of given societies, the actual conception of diseases cannot but reflect the state of the technology, the social

expectations, the division of labor, and the environmental conditions of those populations ([30], p. 252).

Furthermore,

the ascription of "natural" functions to organisms . . . cannot be straightforwardly made on the basis of some empirical inspection of the essential nature of such creatures ([30], p. 249; italics deleted).

Margolis' sociological point in the first passage is well-taken. It seems to me, however, that he slights the distinction between the concept of disease and decisions about the instantiation of the concept. Margolis' argument, moreover, does not touch Boorse's basic thesis, that even though we might have various purposes in life, a healthy body is just one that stays alive sufficiently to permit us to fulfill whatever purposes we do have.

Similarly, H. T. Engelhardt concludes that judgments of health depend on "judgments as to what members of [the] species should be able to do," on our "esteeming a particular type of function" ([14], p. 266), and on "human goals and values" (p. 257). But when Engelhardt writes that one cannot "decide whether a particular state of affairs is *sub specie aeternitatis* functional or dysfunctional" ([14], p. 265) and that there is 'no absolute standard' of health (p. 263), he misconstrues Boorse as aiming at such a standard. The idea that disease is deviation from an evolutionary natural design is clearly not committed to an eternal, changeless catalogue of diseases. Engelhardt goes astray when he argues that because health is understood in terms of human goals (p. 257), all health judgments are evaluative. Boorse defines health in terms of the goal of survival, and because this is a *natural* goal of any organism, the reference to it in picking out diseases does not involve an evaluation.

We have yet to see whether Boorse's account handles 'healthy sexuality' in a satisfactory way. Boorse briefly argues that on his account, homosexuality is a disease because "one normal function of sexual desire is to promote reproduction" ([6], p. 62). Although it is a *disease* – a dysfunctional deviation from the species-design – homosexuality is not necessarily also an *illness* – a disbeneficial disease ([6], p. 61). Boorse's account, apparently, does entail:

(1) homosexuality is unnatural in that it does not conform to the species-typical design, and

(2) homosexuality is a disease because it interferes with the natural function of sexuality (reproduction).

One could challenge Boorse here by arguing that even if conditions that interfere with survival are diseases, interferences with reproduction are not, from the point of view of the individual; such an argument is plausible if we do want to restrict ourselves to the narrow non-evaluative sense of 'healthy body.' Michael Ruse takes a different route. He points out that recent developments in genetics and sociobiology make it possible to deny both (1) and (2), and to understand homosexuality as part of the species-design (a 'balanced heterozygote fitness' theory), or even as "a biologically adaptive move to *increase* one's reproduction in the face of" other disabilities (a 'kin selection' theory) ([41], pp. 715–716; see Engelhardt, [14], p. 265). Notice that Ruse is contending only that Boorse has not necessarily applied his concept of disease correctly to the case of homosexuality. This makes it possible for Boorse to respond in the following imaginary way: "Fine. Ruse's point *confirms* my analysis of health. Now let the scientists decide empirically whether homosexuality is genetic, and if so, what mechanism generates it. As I said, health judgments are a matter for the natural sciences."

Boorse, however, runs into pretty obvious trouble here. The decision as to whether homosexuality is primarily genetic in origin and therefore part of the species-design and not a disease, or primarily environmentally caused and therefore a disease, the result of an external interference with a natural process ([6], p. 59; [8], p. 566), might appear to be a purely scientific matter of sorting out the causes. Yet we also know that this kind of nature/nurture dispute has never been resolved satisfactorily with respect to any socially or politically important property (e.g., racial variations in intelligence). The area, in fact, is quite a mess. In principle (Boorse's words; [7], p. 63), perhaps, there is a scientific solution, but in practice it is elusive. Or rather, what solution is reached is a political matter.

Ruse argues that Boorse is not forced to conclude that homosexuality is a disease, because sociobiological mechanisms may explain its existence. But there is another reason Boorse is not forced to the conclusion that homosexuality is a *sexual* disease. Suppose Boorse is right in saying that homosexuality is a disease in so far as it interferes with reproduction. Then the most we can conclude is that homosexuality is a *reproductive* disease, not that it is a *sexual* disease. (Of course, homosexuality would not even be a reproductive disease for any homosexual who reproduced.) Boorse's reference to the normal function of organs fails for the sex organs, which have several distinct functions (unlike the spleen). Even if Boorse is right that "one normal function of sexual desire is to promote reproduction" ([6], p. 62), there are other functions, not the least of which is to promote pleasure for the organism and thereby to make a contribution to its well-being – perhaps, even, to its survival, if that depends on experiencing pleasure. (See Gray in [44], pp. 169–189, for an account of natural evolutionary sexuality, and Levy's critique, [44], pp. 174–175.) Homosexuality, then, is more like infertility than a sexual deviation (if there are any) like coprophilia; it is not a malfunctioning of the sexual desire/sexual pleasure system, but of reproductive capability. When we make the distinction between reproductive health and sexual health, and focus on the function of the sex organs to produce pleasure, we find the grain of truth underlying the Masters and Johnson approach: sexual dysfunctions are those conditions that interfere with the attainment of sexual pleasure. On this score the homosexual, merely in virtue of being homosexual, does not suffer a sexual dysfunction; indeed, homosexuals who continue to attempt heterosexual relations in the absence of pleasure are sexually dysfunctioning if they could be attaining sexual pleasure homosexually.

On Boorse's view, it appears, the desire to use, or using contraceptives to guarantee the failure of conception, would also count as a reproductive disease, but not a sexual disease; and both homosexuality and the use of contraceptives would not necessarily be reproductive *illnesses*, that is, undesirable reproductive diseases. Notice the difference between Boorse's implied account of contraception and the Roman Catholic view; for both, contraception is unnatural, contrary to the natural design, but for the former this fact yields a disease judgment, while for the latter it yields a moral judgment. On the other hand, Boorse, were he to take into account the sort of sociobiology that Ruse appeals to, could argue that not all contraceptive use is a disease, if at certain times the selective use of contraceptives (even deliberate homosexuality, vasectomy, and sterilization) actually promotes individual survival and reproduction.

There is, of course, some difference between the reproductive disease of infertility and the reproductive disease of homosexuality. Infertility is a physical reproductive disease, while homosexuality, constituted by a type of desire, is a 'mental' reproductive disease, a dysfunction of the affective-emotional system. In infertility the disordered system is the gonads; in homosexuality the main seat of the disorder is the mental apparatus, with the sex organs quite secondary. Boorse does claim that if homosexuality is a disease, it is a mental disease.

For Boorse, "mental health [is] the special case obtained by focusing on the functions of mental processes; and so there is such a thing as mental health if there are mental functions" ([7], p. 63); "one of the most interesting features of the analysis is that it applies without alteration to mental health as long as there are standard mental functions" ([6], p. 58). Of course, we wonder how value-free the specification of "standard mental functions" can be. In one passage Boorse makes an attempt:

It is easy to draw some of the outlines of human mental functions. Perceptual processing, intelligence, and memory clearly serve to provide information about the world that can guide effective action. Drives serve to motivate it. Anxiety and pain function as signals of danger, language as a device for cultural co-operation and enrichment, and so on. If these and other mental processes play standard functional roles throughout the species, we seem to have everything requisite for the possibility of mental health ([7] , p. 64).

When standard mental functions are understood in such a narrow way, Boorse is probably right that he has delineated a universal concept of human mental health. It is difficult to imagine any human being for which, or any human society in which, such a narrowly construed concept did not apply. The cost of this success is that the most interesting questions about mental health – and in particular questions about sexual mental health – are either unanswered or are answered rather trivially with 'not diseases.' For example, if the capacity to experience pain is a standard mental function, as Boorse claims, then it is plausible that the capacity to experience pleasure (e.g., sexual) is also a standard mental function – but this seems to allow *any* sexuality to be healthy, as long as it provides pleasure. One might try to argue that homosexuality is a perceptual dysfunction: one is aroused by the perception (sight, smell) of the wrong objects, or one perceives as desirable objects one should not be perceiving as desirable. But in order to speak about particular diseased perceptions one must go well beyond the bare account offered. One could just as easily say that the heterosexual also has a perceptual disorder: too little of the world is sexually charged, same-sex objects being excluded.

Understandably, Boorse turns to the idea of an empirical developmental psychology to fill in the details of his account of mental health:

> It is an empirical fact that the usual course of human development shows a growth in knowledge of self and world, informed self-acceptance and sense of identity, unification of life goals, tolerance for stress and various kinds of environmental mastery. At any rate, adults tend to have more of these qualities than children. To whatever extent the increase in these traits can be shown part of a normal developmental sequence, it is correct to call them requirements of health in an adult. . . . [T]he mere idea of a developmental disease, i.e., arrest or retardation of normal growth, is not a [confusion]. . . . It is not controversial that a failure to traverse normal stages is unhealthy ([7] , p. 80).

It is not controversial because the claim is an analytic truth.[8] The idea of a developmental psychology is not incoherent, but the world has yet to see any such psychology that is value-free, not even the work of Lawrence Kohlberg [23]. The standard objection here is that Boorse's description of the mature person is a picture of a white, middle-class male, in which case Boorse, like Bentham and Rawls, has hardly avoided evaluations. Furthermore, regarding sexual development, the approach is futile. Very quickly we run into the embarrassing question, 'How much sexual repression, in the Freudian sense, is species-typical in the human design?' Nor does a developmental psychology tell us whether homosexuality deviates from normal development. Clearly, we are back in the muddy waters of the nature/nurture dispute. One very popular solution to these problems will be discussed below (pp. 129–134): natural, and therefore healthy, development is just the development of free persons.

THE SOCIAL DETERMINATION OF HEALTH JUDGMENTS

> . . . health, unlike traffic, cannot be defined, or, rather, is the setting of a continual struggle over definition, a struggle arising from the need to gain control over the realm of the transhistorical, i.e., nature, or the body. . .
>
> Joel Kovel ([24], p. 171)

Whether Levin's (for example) conditions for healthy oral-genital sex are medical criteria, or criteria incorporated from some other discipline (moral philosophy, theology, psychology) and disguised as medical, may not seem to be an important question. After all, what we want to know is whether the criteria, no matter who pronounces them, hold up to critical review. But this is too quick. The 'critical review' is carried out by some discipline and will thereby reflect the interests of yet another professional group. Whether this or that assertion 'belongs' to a discipline is not decided in advance or in the abstract by impartial assignment. Not even philosophy, either as queen or handmaiden, can be counted on. The disciplines fight it out among themselves for control over domains of discourse and their accompanying domains of human behavior, and if medicine is victorious in this struggle, then Levin's conditions *are* medical. The power of the professions in our society (the medicine man, the lawyer, the manager, the priest) is such that gaining effective control over sets of judgments is no innocuous affair.

Ordinarily, the social change in which moral-religious judgments about sexuality were replaced by medical judgments is seen as part of the progress of the humanistic modern age, as the victory of rationality over superstition. Further progress, it is often said, will be made when judgments of sexual sickness are replaced in turn by an acknowledgement of differences, of a plurality of alternative life-styles, thereby fulfilling Mill's values of individuality and autonomous choice.[9] We can already see this trend in Green's client-centered therapy. But this interpretation provides only one side of the story. Rather than interpret these changes as advances in the realm of ideas, or as the moral progress of Western civilization, one could emphasize the competition among prominent professional groups for the social right to incorporate the control of various human behaviors into their spheres of influence. The medical imperialism ([33], pp. 10–11) that grabbed control over sexuality also grabs for control over narcotics addiction, murder, rebellious teenagers, and prostitution (see Note 3). Despite the recent visibility of the 'moral majority' type of organization, the conservative-religious groups have been fighting a losing battle with the moderate-medical establishment and the liberal-professionals over the control of sexuality. We might even expect that the liberal-professional group, with its emphasis on client-centered psychotherapy, will win out simply because individuals are more likely to spend their money to hear that their conditions are acceptable, as long as the client thinks they are acceptable. Thus Green urges a consumer boycott by homosexuals of psychiatrists who label homosexuality an illness, and recommends that they "shop elsewhere for their health, education, and welfare . . ." ([21], p. 348).

There is no transcendent sexual criterion that tells us which judgments (moral/immoral, healthy/unhealthy, criminal/legal, different/same) ought to prevail because they are morally or philosophically superior. There is no transcendent rule that tells us that narcotics addiction ought to be handled by medicine and not by the courts, or vice versa.[10] There is no transcendent criterion, as if everything could be sorted out with an *a priori* logical neatness. Any such criterion is itself a product of a competition among the same social forces. The judgments that prevail, whether low-level judgments about what constitutes healthy sexuality, or high-level judgments about what groups have the right to make judgments about sexuality, are just those that are advanced by the group successful in asserting its power over the others.

"HEALTHY SEXUALITY" AS A POLITICAL CONCEPT

> When he *assigns* sick-status to a client, the contemporary physician might indeed be acting in some ways similar to the sorcerer or the elder; but in belonging also to a scientific profession that *invents* the categories it assigns when consulting, the modern physician is totally unlike the healer. Medicine men engaged in the occupation of curing and exercised the art of distinguishing evil spirits from each other. They were not professionals and had no power to invent new devils. Enabling professions in their annual assemblies create the sick-roles they assign.
>
> Ivan Illich ([22], p. 119)

There were no sexual perverts before 1840.[11] The concept of a 'sexual perversion' came into being (and along with it sexual perverts as an identifiable group of people) at the time sexuality was being thought about as an instinct that was separable from its objects and which, therefore, could latch on to the 'wrong' ones.[12] In creating sexual perversion, medicine created the concept of 'healthy sexuality,' which has remained primarily under its control in a professional 'possession is nine-tenths of the law.' Nowadays, other disciplines, if they are to speak out on healthy sexuality, demur to the primacy of medicine, or cast their criteria in medical terms, or give the appearance of being medical in orientation.

Susan Edwards argues that

> The clinical discourse of the nineteenth century generated a new set of diseases that had hitherto been defined in moral terms. . . . The result of this medicalization of all forms of human activity was to create a definition of normality . . . supported by medical and clinical categories. . . . The disease model succeeded in intervening at every possible level of sexual behavior ([13], pp. 80–81).

Medicine fought with a religious establishment that was already declining in disarray, and appropriated the social right to pass judgments about sexuality. Because medicine justified its judgments in terms of a science of the body, sexual problems had to be analyzed, in the case of women, as gynecological. The "medico-gynaecological model was invoked in order to explain, diagnose, and treat various aberrant expressions of behavior, where in fact, in the majority of cases, no evidence of organic disease or disorder was present" ([13], p. 75). I would go beyond Edwards. It is not so much to the point to insist that 'in fact' there were no 'organic' disorders (that, for example, 'nymphomania' was not caused by any malfunctioning of the gonads). Rather, the point is that due to the flexibility of 'organic' (recall the contrast between positions (1) and (2), pp. 112–113) and the open-ended nature of any scientific investigation, it is always possible for medicine to defend the heuristic value of believing that an organic basis *can* be found, even if any particular causal chain is not currently supported by empirical data. In this way, medical science creates disorders. Because medicine frequently only offers a promissory note that an organic explanation will be found for a condition, doubts can be raised about whether medicine has truly made an advance over earlier demon theories. At the same time, medicine's occasionally paying off its promissory notes accounts for our willingness to accept its designated disease states even in the absence of established organic causes. Our hopes lead us to be accomplices.

In creating sexual perversion, medicine used a common two-step intellectual procedure. In the first step a conceptual distinction is made between *X* and *not-X*; in the second step, the concepts of *X* and *not-X* are given content by assigning concrete items to these categories.[13] The first step is ontological; the second is what I earlier called 'instantiation.' Thus, medicine distinguishes between 'healthy' sexuality and 'unhealthy' (or 'perverted') sexuality, and then assigns different sexual behaviors or preferences to the first category or the second (see [27] for another example: intelligence). Most debates within the tradition are arguments over the second step: which items should be classified in one way rather than another. The history of the psychiatric classification of homosexuality shows that tensions within medicine, and exchanges among medicine, other professions, and the public, have influenced the assignment of items to the categories but have not appreciably modified the categories themselves (see [29] and [45]).

The conceptual isolation of a distinct sexual instinct – an instinct that pushes an organism to experience somatic pleasure, not an instinct pushing people to make babies – made it possible to understand sexuality developmentally, and to ask interesting questions: why do some men prefer brunettes to blondes, or small-breasted women to large-breasted; why do some people enjoy sex with the other gender, and some with their own gender? Postulating an object-neutral pleasure instinct permitted studying the developmental process by which the instinct does latch on to particular objects, and the social influences on this development. The isolation of the instinct from the various acts and objects that provide somatic pleasure, however, makes the giving of concrete content to the two categories difficult. For if the instinct is not necessarily tied to any object, then the separation of acts, objects, and preferences into healthy and unhealthy will look arbitrary. In particular, the distinction between an influence on development that is merely 'formative' (e.g., the influence that creates a preference for brunettes) and one that is 'distorting' (e.g., one which creates coprophilia) will be hard to make out.[14] Even suggesting these examples is problematic.

The solution was to equate the *healthy* development of the sexual instinct with the *natural* development of the instinct, where natural development had to be understood in a special way. 'Natural' development could not just *mean* development toward reproductively effective sexual behavior, and it could not just *mean* 'heterosexual.' That would have upset the research program of eventually providing a successful explanation for the empirical fact that (so much) human sexuality was reproductive and heterosexual. Rather, a more sophisticated notion of the natural was required:

> Natural sexual development is that development persons would undergo in the absence of social influences on their sexual development; natural sexuality is that sexuality persons would exhibit were they never influenced by social factors.

This conception of natural human sexuality will be recognized as derived from liberal state-of-nature political theory, although it can be found also in the non-contractarian social theory of John Stuart Mill:

> I consider it presumption in any one to pretend to decide what women are or are not, can or cannot be, by natural constitution. They have always hitherto been kept, as far as regards spontaneous development, in so unnatural a state, that their nature cannot but have been greatly distorted. . . . This will be less and less the case, but it will remain true to a great extent, as long as social institutions do not admit the same free development of . . . women which is possible to men ([34], pp. 153, 190).

The doctrine underlies the sexual theorizing of Freud[15] and Wilhelm Reich, and anyone else who subscribes to an instinct or quasi-instinct theory of human sexuality. The basic idea is that the instinct, if left alone, will develop in a certain direction; if persons are *free*, if they are not under the control of laws, parental domination, oppressive religious and educational institutions, their sexuality will develop according to the inner laws of the instinct itself, undistorted, unperverted, and healthy. We can thank Freud for what has become a standard piece of contemporary liberal-professional pop psychology.

We should not lose sight of the brilliance of this conception of natural human sexuality. It does not by itself entail that any given act, object, or preference is unnatural and therefore unhealthy and perverted. In one sense (intensionally), it does tell us exactly what natural human sexuality is, but in another sense (extensionally) it doesn't tell us at all what items to assign to the two categories. That move requires further argument, evidence that an item is natural because it would be practiced were there (contrary to fact) no social influences operating on persons. Insofar as further argument is required to establish that a particular behavior is part of the natural developmental sequence, heterosexuality as a standard requires defense and cannot simply be assumed. This made sexual theorizing intellectually challenging; psychiatry could engage in puzzle-solving. The doctrine does, however, suggest a nifty methodology: find natural (i.e., free) persons, study their sexuality, and thereby fill in the content of the categories empirically. Shiploads of European anthropologists thought the Holy Grail would be found in the South Seas.

The doctrine of the equivalence of the natural and the free was also a piece of political ideology that permitted early 20th-century psychiatry to align itself with mainstream liberal economic ideology. The conception of sexual health as the sexually free was a fitting complement of the notion that political and economic well-being is possible only when the capital-owning class is unencumbered by government. (Talk nowadays of a 'healthy' economy sounds like nineteenth century *laissez faire* philosophy; see [4], p. 32.) Healthy firms or persons are not only productive firms, or persons who can perform sexually; firms – and people – can produce most efficiently when they are left alone, free from governmental restraint and interference. In the economic domain, the invisible hand co-ordinates economic behavior; in the sexual domain, there is Natural Selection, and Sexual Selection, to do the job.

But the pride of this approach to 'healthy sexuality,' the fact that further argument is required to carry out the second step, is also its downfall. The difficulty of establishing that *any* sexual act, object, or preference is natural, would be chosen or practiced in a situation of complete freedom, accounts for the history of the endless debates about what is healthy sexuality.[16] Rather than seeing psychiatry as attempting to solve Kuhnian puzzles in a normal-scientific tradition, it is more accurate to see it as bogged down in the swamps of pre-paradigm science. Even though the intention was to assign items to the two categories in a non-arbitrary, empirical, and non-evaluative way, the only recourse was to incorporate other kinds of judgments: not only moral, but also aesthetic and prudential values. At its worst, what was originally presented as an empirical account of healthy human sexuality became a masquerade for factional views as to what sexual behaviors are proper, decent, morally right, permissible, obligatory, attractive, repulsive, or contrary to self-interest. Or the content of the categories was filled in one-dimensionally by referring to simple statistical regularity.

Now, the problems in filling in the content of the categories 'healthy' and 'unhealthy' sexuality can be avoided by abandoning that step, and letting the first step do all the work, or at least as much work as it can do. 'Healthy sexuality' is just *whatever* it is that is chosen by free people, and no one need say anything more than this. (This is quite different from Engelhardt's suggestion that the healthy is that which maintains our freedom; [14], p. 266.) In this case the importance of medicine vanishes, and the issue is seen to be the political issue over freedom which it had always been, but was hidden from view. It follows from the bare analysis of 'health' that it would be contradictory to say of any particular case that an informed and freely-chosen sexual act was unhealthy. This 'radical' idea is consistent with the emerging emphasis on Millian values. Perhaps the goal will become to generate as much personal freedom as possible, and to let sexuality take whatever course it takes under those circumstances. In the meantime, we should not be (and cannot be) so concerned with the question of distinguishing healthy from unhealthy sexuality; a relaxed lack of concern might be a nice antithesis to our current obsession with sex and sexual matters.

Abandoning the second step only appears to avoid evaluations. The evaluative grounding of this concept of sexual health occurs right in the first step, and is not only due to the historical accidents of how the second step is carried out in the face of epistemological calamity. For when it is claimed that healthy sexuality is the sexuality of free persons, a useful and precise concept of 'healthy sexuality' is no longer a biological but a full-blown value concept. It is not the *analysis* of the healthy as the free that invokes an evaluation; that bare analysis commits no one to anything interesting. But to make the analysis useful and precise it is necessary to explain what is meant by 'free.' What we mean by 'free' is itself a matter of moral deliberation, insofar as it presupposes a view of human nature, its capabilities, and a picture of the ideal. What is freedom to the anarchist is not freedom to the liberal, or to the Marxist,[17] or to the conservative, and there are terrible disputes over what social, economic, and political conditions are required for full human freedom. 'Healthy sexuality' turns out to be a notion essentially derived from political philosophy; it should come as no surprise, then, that various social forces fight for control of this concept and its application to the lives of people. It grows out of the political, and it has political ramifications.

University of New Orleans,
New Orleans, Louisiana,
U.S.A.

NOTES

[1] "When I say that mental illness is a myth or that it does not exist, I do not mean to deny the reality of the phenomena to which the term is applied. Human misery and unhappiness exist; conflict and violence exist; inhibited sexual and social behaviors exist. . . . When I say that mental illness is a myth, I am simply asserting that these problems are neither mental nor medical" (Szasz, [46], p. vii).

[2] For the outlines of a materialist psychic science that preserves the autonomy of psychiatry, see Boorse [7].

[3] The editors of the journal *Medical Aspects of Human Sexuality* (a title suggesting, contrary to the wide-ranging content of the journal, that there could be non-medical aspects of human sexuality), claim that "sex-related problems are the proper concern of every physician" [2]. Consistently, this issue of the journal includes articles on the decriminalization of prostitution, coital frequency, and the work ethic. I suppose the reason medicine is concerned with the prostitution question is that certain proposals for decriminalization would put physicians in yet another driver's seat, by requiring periodic medical examinations. It is interesting that the birth control movement did not succeed until the 1920s, when it became "a crusade of a professional medical elite" ([43], p. 18). Did physicians foresee that by supporting birth control they would be able to include it in their domain and thereby enhance their income, importance, and power?

[4] The whole inspiration of Burt and Meeks' *Toward a Healthy Sexuality* is the 'humanistic' idea that healthy sexuality is informed by love. As if announcing a major theoretical

accomplishment, they write: "It was finally the twentieth century that bravely claimed that sex was for pleasure as well as reproduction and that both aspects belonged within a relationship based on love" ([10], p. 8). This is the sort of bravery applauded by the Vatican. But at least Burk and Meeks do not attempt to pass off their wisdom as medical truth.

[5] That Marmor condemns indiscriminatory sex is surprising, given his tolerance of homosexuality ([21], p. 349) and his claim that psychiatrists "do not have the right to label behavior that is deviant from that currently favored by the majority as evidence per se of psychopathology" ([45], p. 1208).

[6] It is of course oversimplified to say that homosexuals cannot engage in heterosexual activity effectively. Many homosexuals, having experienced both, prefer one kind of sexual activity to another. Socarides is wrong to think that homosexuals want to engage in ('turn to') their sexual activity *because* they are incompetent heterosexuals.

[7] In the spirit of a certain kind of sexual egalitarianism, the authors of [19], all women, do not apply 'performance ability' only to men, but also to women. It used to be common, and perhaps still is, to think that only men had to be concerned with performance, the ability to erect and to screw for as long as necessary or desired, while women's sexual task and role was to be performed upon. But the extension of 'performance' to women by the authors does *not* indicate that they have in mind the more appealing sort of sexual egalitarianism that encourages women to be sexually active and aggressive rather than sexually passive. Rather, their emphasis on orgasmicity in women shows that women's sexual performance is merely to be quantified in the same old way in which men's sexual performance had been quantified – by counting.

[8] Theodore Mischel criticizes Boorse, but eventually seems to agree with him that "if evaluations of psychological maturity and immaturity can be grounded in an empirical theory about the course of personality development and its vicissitudes," then "such claims can be justified objectively" ([35], p. 209).

[9] It is consistent with Sedgwick's prediction that more conditions will be considered illnesses, as medicine continues to make *biological* progress ([42], p. 37), that we will recognize fewer specifically sexual illnesses.

[10] Veatch tries, but doesn't come up with much, when he writes: "If the question is to determine whether a deviancy is in one model or another, the methodology must involve, as a minimum, not only the acceptance of the deviancy by the experts of one model, but also the rejection of the deviancy by the experts of all other relevant models" ([48], p. 73). My point is that such 'acceptance' and 'rejection' is the *result* of the competitive struggle among disciplines that call themselves 'relevant,' and so cannot be a methodology.

[11] Michel Foucault suggests that the publication of Heinrich Kaan's *Psychopathia Sexualis* in 1846 marks the birth of the medicalization of the sexual: "the opening up of the great medicopsychological domain of the 'perversions' . . . was destined to take over from the old moral categories of debauchery and excess" ([18], p. 118; see also p. 43, on the birth of medical homosexuality in 1870). Gold ([45], pp. 1211–1212) provides a light version of the position that medicine creates diseases, in particular homosexuality.

[12] "[Medicine] . . . isolated a sexual 'instinct' capable of presenting constitutive anomalies, acquired derivations, infirmities, or pathological processes. . . . The medicine of perversions and the programs of eugenics were the two great innovations in the technology of sex . . ." (Foucault, [18], pp. 117–118).

[13] In creating diseases by formulating concepts, medicine provides many counterexamples to the standard philosophical wisdom, expressed recently by Richard Rorty, that it is "absurd," and "almost no one wishes to say" that "we make objects by using words" ([40], p. 276).

[14] See Nagel ([37], pp. 49–51), and Brown ([9], p. 19), who relies on this point in criticizing Boorse.

[15] For example: "Can an anthropologist give the cranial index of a people whose custom it is to deform their children's heads by bandaging them round from their earliest years? . . . So long as a person's early years are influenced not only by a sexual inhibition of thought but also by a religious inhibition and by a loyal inhibition derived from this, we cannot really tell what in fact he is like" ([20], pp. 47–48).

[16] David Boadella seems to forget the epistemological problems in determining what people would do were they free, when he confidently writes: "Nothing angers the champions of an indiscriminate sexual freedom more than the concept of 'healthy sexuality'. . . . Reich's concept of a healthy sex-economy offered . . . a precise, definable and discriminating goal for the sexual freedom movement . . . " ([5], p. 147). Reich's 'precise' goal was heterosexual genitality, which stands on no firmer epistemological foot than any other conception of natural human sexuality. The idea that natural sexuality is the sexuality of free persons, unencumbered by society, allows even the following sort of shenanigan, a soliloquy of a fetishist imagined for us by Morse Peckham ([38], p. 44): "What people find sexually stimulating is a function of the culture they have been trained in. . . . I have transcended the limitations of my culture; I have been imaginative and creative; my powerful genital response to drawings of black leather corsets is proof that I have a freedom which ordinary men do not have."

[17] There are also those – mostly Marxists – who would say that healthy sexuality is just the sexuality of free persons, without also claiming that this sexuality is 'natural.' This is one way to read Engels: "But what will there be new? That will be answered when a new generation has grown up: a generation of men who never in their lives have known what it is to buy a woman's surrender . . .; a generation of women who have never known what it is . . . to refuse to give themselves to their lover from fear of economic consequences. When these people are in the world . . . they will make their own practice . . . and that will be the end of it" ([16], p. 145).

BIBLIOGRAPHY

1. ______:1972, 'The Porn Industry', *British Medical Journal* #5830 (30 September), 779.
2. ______: 1974, 'A Statement of Purpose', *Medical Aspects of Human Sexuality* **8** (#4).
3. ______: 1981, 'Constitution of the World Health Organization', in [12], pp. 83–84.
4. Barrett, W.: 1978, Contribution to 'Capitalism, Socialism, and Democracy. A Symposium', *Commentary* **65** (#4), 29–71, at 31–33.
5. Boadella, D.: 1973, 'Pseudo-Sexuality and the Sexual Revolution', in D. Holbrook (ed.), *The Case Against Pornography*, Open Court, LaSalle, IL, pp. 144–156.
6. Boorse, C.: 1975, 'On the Distinction Between Disease and Illness', *Philosophy and Public Affairs* **5** (#1), 49–68.

7. Boorse. C.: 1976, 'What a Theory of Mental Health Should Be', *Journal for the Theory of Social Behavior* **6** (#1), 61–84.
8. Boorse, C.: 1977, 'Health as a Theoretical Concept', *Philosophy of Science* **44** (#4), 542–573.
9. Brown, R.: 1977, 'Physical Illness and Mental Health', *Philosophy and Public Affairs* **7** (#1), 17–38.
10. Burt, J. J. and Meeks, L. B.: 1973, *Toward a Healthy Sexuality*, Saunders, Philadelphia.
11. Callahan, D.: 1973, 'The WHO Definition of "Health"', *Hastings Center Studies* **1** (#3), 77–87.
12. Caplan, A. L. *et al.* (eds.): 1981, *Concepts of Health and Disease*, Addison-Wesley, Reading, Massachusetts.
13. Edwards, S.: 1981, *Female Sexuality and the Law*, Martin Robertson, Oxford, England.
14. Engelhardt, H. T., Jr.: 1976, 'Ideology and Etiology', *Journal of Medicine and Philosophy* **1** (#3), 256–268.
15. Engelhardt, H. T., Jr.: 1981, 'Human Well-being and Medicine: Some Basic Value-judgments in the Biomedical Sciences', in T. Mappes and J. Zembaty (eds.), *Biomedical Ethics*, McGraw-Hill, New York, pp. 213–222.
16. Engels, F.: 1972, *The Origin of the Family, Private Property and the State*, International Publishers, New York.
17. Feinberg, J.: 1970, 'Crime, Clutchability, and Individual Treatment', in *Doing and Deserving*, Princeton University Press, Princeton, pp. 252–271.
18. Foucault, M.: 1976, *The History of Sexuality*, Vol. 1, Vintage, New York.
19. Frank, E. *et al.*: 1978, 'Frequency of Sexual Dysfunction in "Normal" Couples', *New England Journal of Medicine* **299** (#3), 111–115.
20. Freud, S.: 1961, *The Future of an Illusion*, Norton, New York.
21. Green, R.: 1981, 'Homosexuality as a Mental Illness', in [12], pp. 333–351.
22. Illich, I.: 1976, *Medical Nemesis*, Pantheon, New York.
23. Kohlberg, L.: 1981, *The Philosophy of Moral Development*, Harper and Row, New York.
24. Kovel, J.: 1982, *The Age of Desire*, Pantheon, New York.
25. Lasch, C.: 1979, *The Culture of Narcissism*, Norton, New York.
26. Levin, M.: 1973, 'Pornography and Redeeming Social Values', *Current Medical Dialogue* **40**, 621, 624.
27. Lewontin, R. C.: 1982, Letter to *The New York Review of Books* (4 February), 40–41.
28. Lumiere, R. and Cook, S.: 1983, *Healthy Sexuality and Keeping It That Way*, Simon and Schuster, New York.
29. Margolis, J.: 1975, 'The Question of Homosexuality', in R. Baker and F. Elliston (eds.), *Philosophy and Sex*, Prometheus, Buffalo, NY, pp. 288–302.
30. Margolis, J.: 1976, 'The Concept of Disease', *Journal of Medicine and Philosophy* **1** (#3), 238–255.
31. Marmor, J. *et al.*: 1977, 'Viewpoints: What Distinguishes "Healthy" from "Sick" Sexual Behavior?', *Medical Aspects of Human Sexuality* **11** (#2), 67, 75–77.
32. Masters, W. and Johnson, V.: 1970, *Human Sexual Inadequacy*, Little, Brown, Boston.

33. Mechanic, D.: 1973, 'Health and Illness in Technological Societies', *Hastings Center Studies* **1** (#3), 7–18.
34. Mill, J. S.: 1970, *The Subjection of Women*, in A. Rossi (ed.), *Essays on Sex Equality*, University of Chicago Press, Chicago, pp. 123–242.
35. Mischel, T.: 1977, 'The Concept of Mental Health and Disease: An Analysis of the Controversy Between Behavioral and Psychodynamic Approaches', *Journal of Medicine and Philosophy* **2** (#3), 197–219.
36. Murray, R. and Callahan, D.: 1974, 'Genetic Disease and Human Health', *Hastings Center Report* **4** (#4), 4–7.
37. Nagel, T.: 1979, 'Sexual Perversion', in *Mortal Questions*, Cambridge University Press, Cambridge, England, pp. 39–52.
38. Peckham, M.: 1969, *Art and Pornography*, University of Chicago Press, Chicago.
39. Rice, F. P.: 1978, *Sexual Problems in Marriage*, Westminster, Philadelphia.
40. Rorty, R.: 1979, *Philosophy and the Mirror of Nature*, Princeton University Press, Princeton, NJ.
41. Ruse, M.: 1981, 'Are Homosexuals Sick?', in [12], pp. 693–723.
42. Sedgwick, P.: 1973, 'Illness – Mental and Otherwise', *Hastings Center Studies* **1** (#3), 19–40.
43. Snitow, A. *et al.* (eds.): 1983, *Powers of Desire*, Monthly Review Press, New York.
44. Soble, A. (ed.): 1980, *The Philosophy of Sex*, Littlefield, Adams, Totowa, NJ.
45. Stoller, R. *et al.*: 1973, 'A Symposium: Should Homosexuality Be in the APA Nomenclature?', *American Journal of Psychiatry* **130** (#11), 1207–1216.
46. Szasz, T.: 1974, *The Ethics of Psychoanalysis*, Basic Books, New York.
47. Veatch, R.: 1973, 'Generalization of Expertise', *Hastings Center Studies* **1** (#2), 29–40.
48. Veatch, R.: 1973, 'The Medical Model: Its Nature and Problems', *Hastings Center Studies* **1** (#3), 59–76.
49. Wiewel, B.: 1981, Letter to *Ms.* (March), 7.

JOSEPH MARGOLIS

CONCEPTS OF DISEASE AND SEXUALITY

One can't help noticing how much the power of Socrates' discussion of virtue in the state and the human soul depends on confusing – perhaps ironically – the tautologically assigned functions of various social professions and roles and the seemingly equally easy identification of the natural function of human beings. In the *Republic*, for instance, quite slyly, Socrates asks Thrasymachos: "Do you think there is a virtue in each thing which has a work appointed to it?" He then mentions the eyes and the ears and somehow springs at once to the work of the soul – "to care, to rule, to plan, and all things like that"; in fact, the virtue of the soul is announced as 'justice', though in closing this first portion of the dialogue Socrates confesses to not even knowing what justice is (*Republic*, Bk. 1).

The discussion had been prepared by indisputable examples of the clear function of a ship's pilot, the good checkers player, the shepherd, the physician – and again, quarrelsomely, the ruler. And what the thoughtful reader keeps his mind peeled for are signs of plausible arguments telling us what *the function of man is*, in virtue of which we could claim an assured grasp of the point of the persuading analogies and in particular of the objectivity of the very norms of medicine and statesmanship. Any attentive reader of the *Dialogues* knows that Plato never satisfies us on this count, and he may (rightly) suspect that no one ever has or ever will. Even Thrasymachos, who is hardly a match for the old man, baits Socrates, observing that shepherds and oxherds don't look "for the good of their sheep and cattle . . . they fatten them for their own good and their masters'." By parity of reasoning, medicine and statesmanship may well serve (Thrasymachos crudely puts it) "the advantage of the stronger," that is (trivially), the implied advantage or supposed advantage of whatever social groups effectively determine the operative norms of health and social justice – because there is no other way to decide the matter. The cynicism is clear enough; but, although it is more usual in the political setting, it is certainly intriguing to consider

E. E. Shelp (ed.), Sexuality and Medicine, Vol. I, 139–152.

the analogy between medicine and statesmanship that way, particularly if we hold the question of sexual health in the back of our minds.

It is extraordinary how quickly people leap to the conclusion that there *must* be some natural, objectively discernible, perhaps even essential norms of health, moral well-being, sexual relations, or even qualification as being of a so-called natural kind. There are many encouragements for the implied *non sequitur*. Also, there is no reason to suppose that the resolution of the question in all of the contexts mentioned should behave in any logically uniform way at all. Certainly, it is a dubious leap to judge that if post*men* and fire*men* and police*men* and crafts*men* have functions – as of course they do because their roles are contingently institutionalized (at least) in the practices of particular societies – then *men*, the "*common denominator* in regard to these functions," ([10], [14], [30]) must also have a similarly assignable (now, perhaps natural or even essential) function, perhaps a function that can be derived merely from examining the meaning of the 'natural kind term' *man*. But there is good reason to think that man (the human being) *has* no function, or has no function that can be naturally discerned: (1) by examining his biological behavior somehow independently of whatever medical, moral, legal, political, religious, or other culturally developed norms he happens to affirm or is assigned within the particular interval of history he occupies; (2) by identifying any relatively rich, detailed norms of any of these cultural sorts genuinely supported by a reasoned consensus holding across the whole of human history; or (3) by a universally compelling analysis of what is entailed by the very concept of being a member or instance of *Homo sapiens*, or a person or a man or woman.

No one has as yet offered a convincing account of any of these sorts. We must be careful to admit that if man, or *Homo sapiens*, has no 'natural' function, it does *not* follow that his sexual organs and processes (or eyes) have no 'natural' function (nor does the reverse follow). It *would* trivially follow that *man* had no sexual function, though even that would not show that there was no medically relevant sense in which sexual functioning could be construed as 'healthy' or 'normal' or conducing to one's health. And certainly, to concede that man has no natural function is not to say that there are no 'natural' constraints on what might be reasonably offered as a system of norms of any of the sorts mentioned; nor, of course, is it to deny that *if* the human *species* is to survive, then there is a plausible (and medically relevant) sense in which a form of 'natural' or 'normal' or 'healthy' functioning must be preserved in the *species* – though hardly (for that reason alone) in every member of the species; nor is it to deny that, if one fair account of sexual health and good sexual relations could be plausibly proposed, there could well be other such accounts compatible with whatever minimal (or 'necessary') natural constraints were admitted in the first, but not themselves compatible with the first or with one another.[1]

The complexity of the bare notion of sexual health is impossible to miss even in these obvious and most preliminary reflections. There are profound difficulties regarding the objectivity of generic medical norms of health and disease, of mental health, of the independence of medical and moral concerns (particularly with regard to sexual matters), of the objectivity of moral norms, of the conceptual equivalence of membership in *Homo sapiens* and of being a person, of the functioning of the human species and the members of the species, and of functions assignable to the species or to members of the species, or to human persons as such, as distinct from modes of functional behavior contingently favored in this society or that for any such referents (cf. [13, 15, 19]).

The strongest case for an objective medicine would, of course, hold: (a) that the members of *Homo sapiens* had an essential and discernible function; (b) that the functions assigned to the organ systems and biological processes of the human body were reasonably demarcated subfunctions *of* the molar functioning of *Homo sapiens*; and (c) that medical concern with human culture, human history, human custom and the like (the functioning of human persons, in short) was entirely restricted to the causes bearing of such phenomena on whatever, regarding the molar function of *Homo sapiens* and the subfunctioning somatic parts and processes of *Homo sapiens*, are delimited as the specific concern of medicine. Clearly, a similarly strong view of an objective morality would require fixing the essential function of human persons and human societies, possibly to include the essential function of *Homo sapiens*, but certainly primarily focused on the concept of a person.

There are, in fact, hints here and there in contemporary theories of health and disease that a medically strict conception of disease along the lines of (a)–(c) could be formulated. One well-known physician, Horacio Fabrega, for instance, contrasts what he calls the 'biologistic framework' with a so-called 'sociobehavioral framework' – by which he means to claim that it *is* possible to specify the somatic functions of the human body (bracketing psychiatry and behavioral medicine as frankly 'problematic'), in a way that is 'universal or transcultural' and altogether 'logically independent' of any of the 'social categories' of the alternative framework of disease that understandably evolve within a concerned society [6]. But Fabrega, who has pursued the issue for many years, does not actually provide the required account. In fact, he remarks, quite innocently, that "There is no reason to doubt that man everywhere is biologically the same individual in most of the important respects that enable physicians to make judgments about disease" [6].

The slightest glance, however, at the literature regarding the widespread occurrence and subtlety of sexual anomaly – for example, in the pioneering work of John Money – pointedly shows that Fabrega is just wrong about the biological facts that might have supported the kind of medical essentialism he favors; and what is more, they show unquestionably how, particularly regarding sexual functioning, it is quite impossible to segregate the 'biologistic' factors from the 'sociobehavioral' (cf. [22]). Klinefelter's syndrome, for example, is apparently classified as a form of cytogenetic hermaphroditism (that is, without morphologic hermaproditism); but although its population is at risk with regard to a large variety of forms of 'psychopathology,' including 'severe mental retardation,' the various 'sexual psychopathologies' seem to predominate – including "transvestism, transexualism and various manifestations of homosexuality and bisexuality" ([22], p. 33). Now, of course, it is altogether possible that explicit somatic disorders may be linked to such genetic anomalies (for instance, brain-based impairment of psychologic development) – in which case, diseases of sexual process would be on a par with whatever may be claimed for other somatic disorders. But the status of the sexual-*behavior* disorders (homosexuality or bisexuality, say) is by this time a profound embarrassment in the recent history of medical (and psychiatric) classification ([13, 15–19, 26]. Money, who has been as close as any to the medical complexities of sexual identity, usefully observes: "A dichotomous distinction between male and female, with no allowance for a possible spectrum of variation, makes it impossible for the transexual to maintain his or her self-respect. The function of sex reassignment, a new medical procedure, is to provide yet another acceptable way for resolving the transexual-transvestite's lifetime dilemma. Endocrine therapy and conversion surgery, therefore, represent an investigative attempt to match the body with the mind and social role" ([23], pp. 253–254).

The key distinction, here, is quite a nice one. When one speaks of the function of the eye (with respect to fixing the range of diseases of the eye), one cannot fail to consider the widest possible range and characteristic activities of human beings or human persons. We are actually not obliged to determine the function or functions of humans *in* determining what we take to be the normal range of functioning of the eyes: the most comprehensive and detailed study of the actual variety of human patterns of life (of human *functioning*, so to say) confirms at least a narrowly variable minimal function of the eyes as a subfunctioning part of the highly variable range of molar functioning that humans exhibit. What the precise conceptual justification for this extrapolation is is, of course, controversial, but its reasonableness can hardly be doubted. The peculiarity of sexual functioning, however, is precisely that, behaviorally construed, it *is* a form of the *molar* functioning of humans; and otherwise construed (as in Klinefelter's syndrome), it is utterly indifferent to such molar functioning, unless because of accidental causal complications (for example, as in the way anorexia may threaten the hormonal balance of female adolescents). This shows at one and the same time Fabrega's mistake in attempting to segregate 'biologistic' from 'sociobehavioral' functioning and the essentialism that that attempt presupposes, *and* the peculiar conceptual complication that the very notion of sexual disorder imposes on us.

It is, also, quite instructive to find that a typically successful psychiatric program for managing sex therapy – Helen Singer Kaplan's, for example, at the Cornell University School of Medicine – regards "sex therapy [primarily] as a specialized branch of psychotherapy," that is, as addressed to molar agents as such, behaviorally and intentionally ([11], p. xv). "Sex therapy," we are told, "relies heavily for its therapeutic impact on erotic tasks which the *couple* conducts at home" ([11], p. xiv; italics added). The basis (or professional orientation) is clear enough, but its significance needs to be drawn out. Because 'sexual disorders' are apparently separable into 'variations' and 'dysfunctions': the variations include so-called 'deviations and perversions,' but Dr. Kaplan's extraordinarily successful manual explicitly advises that the variations are not "amenable to sex therapy and for this reason they are not discussed in this book" ([11], pp. 247–250; see also [2, 21]). She focuses understandably on impotence, retarded and premature ejaculation in the male, and vaginismus, general female sexual dysfunction (frigidity) and orgastic dysfunction in the female. It is easy to see, of course, that a more adventurous sex therapy could have been formulated along analogous lines for homosexual and bisexual pairs and even for other more complex relationships. But the most important point to consider is that sex therapy (whether or not restricted to 'normal' heterosexual couples) is conceptually dependent on the morally acceptable or tolerated preferences in the society at large. There is no way to free the medical conception of sexual disorder from such a dependence unless some form of molar essentialism can be defended – and then, there is likely to be a convergence between medical *and* moral essentialism bypassing (or pretending to bypass) the historically contingent practices of a given society.

We have now managed to isolate the salient peculiarities of the medical conception of sexual dysfunction or disorder or disease: (1) it concerns molar behavior and intention rather than subfunctioning parts or processes of molar functioning (like the functioning of the eye); (2) the objectivity of a medicine addressed to sexual dysfunction presupposes an objectively assignable function or functions to humans at the molar level (or at least some suitable alternative for assessing modes of human functioning), or it must be frankly construed in purely instrumental terms, that is, as efficiently servicing whatever, independently, are conceded to be the favored forms of human functioning; and (3) any reasonable biological constraints on such functioning (for instance, considerations of survival and viability) that could legitimate the objectivity of such a medicine can only apply to the species as a whole and not, distributively, to each member of the species (in contrast, precisely, with what may be claimed regarding the function of the eye, which *is* distributively relevant). Hence, *if* man as such has no function, then the assignment of a function to the eye, in the medically pertinent sense, precludes any possibility of disjunctively analyzing the 'biologistic' and 'socio-behavioral' frameworks of human functioning; *and that* entails the reasonableness of extending medicine to some portion at least of the concerns of psychiatry and so-called behavioral medicine. Once this is granted, then the profound importance of the sexual concerns of the human race reasonably stalemate any argument that would preclude questions of sexual dysfunction from forming a distinct part of the central professional concern of medicine. To say that, however, is to confirm as well the conceptual linkage between medical and moral or similarly normative concerns. This may be fairly regarded as a part of the complex issue Socrates originally raised.

Some interesting lessons may be drawn from these quite modest considerations. The most obvious, certainly, is that the strongly polarized form of dimorphic gender identity favored in modern societies and strongly supported in Western medicine is not itself a classificatory disjunction of a 'biologistic' or 'objective' sort or of a sort that collects morally neutral notions of behavioral normality or of a sort that permits medical criteria of sexual function and dysfunction to be formulated in morally or other (similar) normatively neutral terms. The complication is multiple. First of all, it is a schematism automatically invoked (for instance, immediately on medical intake) that is as such indifferent to distinctions of genetic, hormonal, morphologic, and role factors. Secondly, it automatically activates and reinforces distinctly stereotyped and polarized norms of sexual behavior – at once proto-medical and proto-moral – for assessing the life style of particular patients. One has only to remember the bad faith and curious near-panic of medical and hospital professionals, recently, regarding widespread refusals to treat any patient apparently suffering from the so-called AIDS syndrome. Or to remember the odd asymmetries and gymnastic theorizing leading to the American Psychiatric Association's revision of the classification of sexual disorders (cf. [13, 15–19]).[2] The routine patterns of gender identification are, by their very insouciance, treated, in medically as well as morally relevant respects, as if they correctly marked the principal, predominant, invariant, essential, typical, or strongly characteristic molar orientation, behavior, and interests of those thus classified – or at least the norms to which they should conform. This means both that dimorphic gender identity somehow is determinative (in some significant degree) of the entire molar orientation of humans and that it is surely determinative of their molar sexual orientation. This is why it is so curious, if one thinks about it, that humans are classified in a medically relevant sense not only as male or female *tout court* but also, given a more detailed history, as homosexual or not homosexual.

Statistical deviance, behavioral or biological, is hardly a sufficient (or at times even relevant) condition for determining medical disorder, dysfunction, disease (or illness); and medical disorder hardly entails statistical deviance. There *is* an ineliminable normative constraint at work in notions of health and disease; and, *if* man cannot be assigned a normative nature or function *qua* man (along the lines already sketched), then the required concept of medically (or morally, or medically-morally, or normatively) pertinent deviance is a 'social construction' of some sort, that is, a model generated on the basis of a review of the behavior of human societies. Robert Veatch, who presses this thesis (which in effect goes contrary to Fabrega's view), manages nevertheless to support a medically (and morally) relevant essentialism even broader than Fabrega's. His principal worry here is to counter relativism ([28, 29]; but cf. [20]). "The medical model," he says, "is a systematic mode of interpretation of a type of social deviance [that] incorporates negative evaluations of the deviancy." Specifically: "A deviancy will be placed within the medical model if it is seen as (a) non-voluntary and (b) organic, if (c) the class of relevant, technically-competent experts is physicians, and if (d) it falls below some socially defined minimal standard of acceptability" ([29], p. 64). There is no need to worry Veatch's definition here. The fact is that Veatch explicitly maintains that "it is our position that there are 'absolute and objective values', [he means to support some form of scientific naturalism] upon which one may properly . . . construct a socio-cultural understanding [of both medicine and morals. But to pursue that issue would, he says] lead into the realm of metaphysics and theology, [perhaps even to] debates about the nature of the transcendent"[3] ([29], p. 61, N. 4). Along these lines, he mentions homosexuality – and maintains: "Homosexuality lends itself to organic interpretation primarily at the causal level with research on andosterone/estrogen ratios suggesting abnormal balances in homosexuals. . . . It appears that the more clearly the deviance is associated with organicity . . . the more nearly it fits the medical model" ([29], p. 70). But what Veatch fails to consider is the inherently circular nature of determining the supposed organic considerations in question: the 'normal' ratios are clearly meant to depend on the dimorphic model favored – which, presumably, they help to fix in the first place; and Veatch's linkage (in his discussion) of the organicity of homosexuality with that of alcoholism and nicotinism is self-explanatory.

The view here favored agrees with an opinion of Christopher Boorse's, developed in a number of well-known papers, namely, that "statistical normality is . . . never necessary nor sufficient for clinical normality ([3], p. 50). But Boorse takes an extreme position, categorically rejecting the thesis "that health is an essentially evaluative notion [as] entirely mistaken," which he says is due to "a confusion between the theoretical and the practical senses of 'health' or in other words, between disease and illness." In this spirit, he adds that he assumes "that the idea of health ought to be analyzed by reference to physiological medicine alone" – allegedly in accord with the practice of medical textbooks; and that "it is a mistake to view physical and mental health as equally well-entrenched species of a single conceptual genus" ([3], pp. 49–50).[4]

Boorse cites approvingly the opinion of C. Daly King: "The normal . . . is objectively, and properly, to be defined as that which functions in accordance with its design" [12]. He explains his own view – and his sense of King's view – as follows: "The crucial element in the idea of a biological design is the notion of a natural function . . . a function in the biologist's sense is nothing but a standard causal contribution to a goal actually pursued by the organism . . . The specifically physiological functions of any [organismic] component are, I think, its species-typical contributions to the apical goals of survival and reproduction" ([3], p. 57). On this basis, Boorse holds that "biological function statements are descriptive rather than normative claims." And in accord with this view, Boorse notes that "Functions are not attributed to this context to the whole organism at all, but only to its parts, and the functions of a part are its causal contributions to empirically given goals" ([3], p. 58). By this strategy, then, he intends to meet the general objection originally drawn from Socrates' discussion regarding the function of man and, at the same time, to divest the medical concept of health of normative import.

There is, however, a critical and fatal equivocation in Boorse's account. He had claimed, as already noted, that functions are assigned not to 'the whole organism' but 'only to its parts.' But he also wished to maintain (and is obliged to maintain, by his intended argument) that "what makes a condition a disease is its derivation from the natural *functional organization of the species*"; "deficiencies in the functional efficiency of the body are diseases when they are unnatural, and they may be unnatural either by being atypical or by being attributable mainly to the action of a hostile environment" ([3], p. 59; italics added). Now, Boorse's medically decisive 'empirically given goals' are survival and reproduction. But survival and reproduction are equivocal as regards the extension of a single organism's life and the viability of an entire species. In particular, what 'typically' is required for the survival and reproduction of the *species* is quite compatible with a variety of sexual behavior patterns – and, more important, in Boorse's terms, with a variety of reproductive capacity (even the distributed absence of such capacity) – with respect to individual organisms. In fact, what is 'typical' for the species is reasonably straightforward, even if what is 'typical' for the aggregated members of the species is quite controversial and unsettled and perhaps even impossible to determine in any 'descriptive' or 'non-normative' sense.

It is *not* important to decide which sense to favor here: the point is that *if* the species-sense is preferred, then the question of the health of the members of that species is *not yet addressed*; and *if* the member-sense is preferred, then the question of how to decide what pertinently is 'typical' for the aggregated members *cannot* be specified (however Boorse may wish to decide the matter) in a normatively neutral way. First of all, the inference from what is typical for the survival and reproduction of the species to what is 'natural' or 'normal' for individual members involves some reasoned decision which is not obviously value-free or logically entailed; and secondly, individual sexual functioning (and capacity) can hardly be said, given the immense variety of gender-identity features that are viable both in terms of individual lives and species survival and reproduction, to conform in any obvious way with what Boorse rather loosely specifies as 'a goal actually pursued by the organism.' It is, therefore, difficult to see in what sense Boorse can maintain his 'descriptive' thesis with regard to sexual functioning. He does say that "most people . . . want to pursue the goals with respect to which physiological functions are isolated . . . [they] want to engage [for instance] in those particular activities, such as eating and sex, by which these goals are typically achieved" ([3], p. 60). But it is not clear whether they want to engage in sexually *typical* activities (whatever, reasonably, may be meant by that) or whether they *typically* want to engage in some of the sexual activities (of the immense variety that the species sustains); and a significant number may *typically* not want to engage in such activities (without its being entirely clear whether, in doing so, they are behaving untypically or what the sense is in which they are behaving typically). Now, the point of fretting these distinctions is *not* to recover psychiatric or behavioral medicine by the back door (though it does strengthen its case, in spite of Boorse and Szasz). It is rather to show that the putatively physiological sense of sexual disorder and disease cannot be neatly linked to the goals of survival and reproduction (certainly cannot be so linked in a normatively neutral sense) *and* cannot really be free from considerations of prevailing social practice. Also, there is reason to think that, for instance with respect to aging, similar difficulties are bound to confront Boorse's thesis.

In any case, there is reason to think that Boorse has somehow conflated the logic of the functioning of the eye and the logic of sexual functioning: it is very difficult to suppose that sexual functioning concerns 'parts' (in Boorse's sense) and not molar organisms, and it is equally difficult to suppose that if it concerns the 'functional organization of the species' (taken collectively), any straightforward conclusions can be drawn about the medically relevant form of sexual health among individual organisms. But it is perfectly obvious that the sexual functioning of individuals *is* a salient concern of professional medicine. And it seems equally obvious that, for all his resistance, Boorse's theory of species-wide regularities regarding survival and reproduction is merely a cryptic (though not at all typical) form of the very normativism he professes to have exposed. There is a basic division of labor, so to say, relevant to species survival that simply does not bear at all or at all directly on the functioning of the individual members of a species.

Finally, it is a very curious fact that the influential doctrine which Pope Paul VI espouses in the encyclical *Humanae Vitae* (1964) not only orients the morally informed sense of medical practice among Roman Catholic physicians and patients, but also gains a great deal of its plausibility by implicitly conflating the putatively essential function of humans as molar agents with the conditions for the functional survival of the species – distributing the latter (under a certain constraint) among the former. Thus, Pope Paul construes explicit sexual acts like intercourse, with or without consummation, solely in terms of the 'conjugal act,' and then construes 'conjugal love' as essential to conjugal acts (hence, performed solely within a licit marriage: 'human,' 'total,' 'faithful,' 'exclusive,' and 'fecund'), "safeguarding both these essential aspects, the *unitive* and the *procreative*" (that is, uniting "husband and wife . . . in chaste intimacy" and invariably "open to the transmission of life").[5] But on the argument advanced, there is no demonstrable basis for affirming the function of men *qua* men (either morally or medically), and the 'natural' functioning of the species (as reproducing itself) is entirely compatible with a wide variety of morally and medically normative schemes.

There is, in short, no reasonable way to preclude models of sexual behavior and intent from playing a salient role within a medical conception of sexual health and disease – if sexual functioning is, as such, deemed medically relevant; and there is no reasonable way to include such models free of the contingent moral and related normative convictions of the particular societies that do attempt to construct for themselves a medical conception of sexual health. Sexual health cannot be restricted to any form of sub-molar functioning; and pertinent medical concerns with species-wide reproduction and survival yield no distributive criteria or principles or norms of sexual health and disease as such. Health is a molar concept, but there is no doctrinally neutral or disinterested sense in which molar functions are assigned the members of natural species or human persons. It is, therefore, pointless to pretend to exclude as important a function as the sexual from the formulation of what, in the medical sense, is taken to constitute a reasonable model of health; but to include it entails the impossibility of treating health as a non-normative concept *and* the impossibility of distinguishing between purely medical and moral concerns. Also, the admission of the sexual in this regard is hardly unique: very much the same consequences follow from the admission of aging and of mental and emotional capacity – although aging is not narrowly associated with the functioning of organ 'parts' or sub-processes (though it could be) and although mental and emotional capacity are, at least within a certain range (of interest to psychiatry, behavioral medicine, and psychotherapy), probably not closely linked with dysfunctional 'parts' or sub-processes.

Temple University,
Philadelphia, Pennsylvania,
U.S.A.

NOTES

[1] The issue depends on the formal viability of a relativistic view. See for instance [20].

[2] For a flavor of 'second-generation' worries about sex-bias in the *DMS-III*, see *American Psychologist* 38, No. 7, July, 1983.

[3] He claims to have pursued these matters in [28] but the discussion there is entirely formal; he offers no evidence supporting an empirical or scientific essentialism.

[4] Boorse professes to follow Antony Flew and Thomas Szasz here. But Flew's argument is utterly indecisive regarding the restriction of general medicine: he is primarily concerned to resist certain excessive tendencies to dismiss certain forms of criminality as 'mental illness.' And Szasz's argument against the conceptual legitimacy of mental illness is simply a profound confusion. See [7]; also [23] and [13]. In fact, in a symposium at Syracuse University, Winter, 1967, at which he responded to a paper of mine criticizing his thesis, Szasz conceded to me privately that, although he did not believe his argument to be theoretically compelling, he did feel that it was important in a 'practical' sense to maintain it – that is, given prevalent, undesirable psychiatric practices.

[5] The text is given, conveniently, in [1], pp. 131–149; italics added.

BIBLIOGRAPHY

1. Baker, R. and Elliston, F. (eds.): 1975, *Philosophy and Sex*, Prometheus Books, New York.
2. Bell, A.: 1975, 'The Homosexual as Patient', in R. Green (ed.), *Human Sexuality: A Health Practitioner's Text*, Williams and Wilkins, Baltimore, pp. 55–72.
3. Boorse, C.: 1975, 'On the Distinction Between Disease and Illness', *Philosophy and Public Affairs* **5**, 49–68.
4. Brody, B. and Engelhardt, H. T., Jr. (eds.): 1980, *Mental Illness: Law and Public Policy*, D. Reidel, Dordrecht, Holland.
5. Caplan, A. *et al.* (eds.): 1981, *Concepts of Health and Disease*, Addison-Wesley, Reading, Mass.
6. Fabrega, H.: 1972, 'Concepts of Disease: Logical Features and Social Implications', *Perspectives in Biology and Medicine* **15**, pp. 583–616.
7. Flew, A.: 1973, *Crime or Disease?*, MacMillan, London.
8. Green, R. (ed.): 1975, *Human Sexuality: A Health Practitioner's Text*, Williams and Wilkins, Baltimore.
9. Green, R. and Money, J. (eds.): 1969, *Transsexualism and Sex Reassignment*, Johns Hopkins University Press, Baltimore.
10. Hampshire, S.: 1959, *Thought and Action*, Chatto and Windus, London.
11. Kaplan, H. S.: 1974, *The New Sex Therapy*, Brunner/Mazel, New York.
12. King, C. D.: 1945, 'The Meaning of Normal', *Yale Journal of Biology and Medicine* **17**, 493–501.
13. Margolis, J.: 1966, *Psychotherapy and Morality*, Random House, New York.
14. Margolis, J.: 1971, *Values and Conduct*, Clarendon Press, Oxford.
15. Margolis, J.: 1975, *Negativities: The Limits of Life*, Charles E. Merrill, Columbus, Ohio.
16. Margolis, J.: 1975, 'The Question of Homosexuality', in R. Baker and F. Elliston (eds.), *Philosophy and Sex*, Prometheus Books, New York, pp. 288–302.
17. Margolis, J.: 1976, 'The Concept of Disease', *Journal of Medicine and Philosophy* **1**, 238–255.
18. Margolis, J.: 1980, 'The Concept of Mental Illness: A Philosophical Examination', in [4], pp. 3–24.
19. Margolis, J.: 1982, 'Homosexuality', in T. Regan and D. Van DeVeer (eds.), *And Justice for All: New Introductory Essays in Ethics and Public Policy*, Rowland and Littlefield, Totowa, New Jersey, pp. 42–63.
20. Margolis, J.: 1983, 'The Nature and Strategies of Relativism', *Mind* **92**, 548–567.
21. Masters, W. H. and Johnson, V. E.: 1979, *Homosexuality in Perspective*, Little, Brown, Boston.
22. Money, J. and Ehrhardt, A. A.: 1972, *Man and Woman; Boy and Girl*, Johns Hopkins University Press, Baltimore.
23. Money, J. and Schwartz, F.: 1969, 'Public Opinion and Social Issues in Transsexualism: A Case Study in Medical Sociology', in R. Green and J. Money (eds.), *Transsexualism and Sex Reassignment*, Johns Hopkins University Press, Baltimore, pp. 253–270.

24. Regan, T. and Van DeVeer, D. (eds.): 1982, *And Justice for All: New Introductory Essays in Ethics and Public Policy*, Rowland and Littlefield, Totowa, New Jersey.
25. Rouse, W. H. D. (trans.): 1956, *Great Dialogues of Plato*, Mentor, New York.
26. Ruse, M.: 1981, 'Are Homosexuals Sick?' in A. Caplan *et al.* (eds.), *Concepts of Health and Disease*, Addison-Wesley, Reading, Mass. pp. 693–723.
27. Szasz, T.: 1961, *The Myth of Mental Illness*, Harper-Hoeber, New York.
28. Veatch, R.: 1973, 'Does Ethics Have an Empirical Basis?', *The Hastings Center Studies* **1**, no. 1, 50–65.
29. Veatch, R.: 1973, 'The Medical Model: Its Nature and Problems', *The Hastings Center Studies* **1**, no. 3, 59–76.
30. Vendler, Z.: 1967, 'The Grammar of Goodness', in Z. Vendler, *Linguistics in Philosophy*, Cornell University Press, Ithaca, New York, pp. 172–195.

JEROME NEU

FREUD AND PERVERSION

The first of Freud's *Three Essays on the Theory of Sexuality* is entitled 'The Sexual Aberrations.' Why should Freud begin a book the main point of which is to argue for the existence of infantile sexuality with a discussion of adult perversions (after all, the existence of the adult aberrations was not news)? While many answers might be suggested with some plausibility (e.g., to ease the shock of the new claim; or, medical texts typically begin with pathology), I think Freud's beginning can be usefully understood as part of a brilliant argumentative strategy to extend the notion of sexuality by showing how extensive it already was. Freud himself (in the Preface to the Fourth Edition) describes the book as an attempt "at enlarging the concept of sexuality" ([11], p. 134). The extension involved in the notion of perversion prepares the way for the extension involved in infantile sexuality.

The book begins, on its very first page, with a statement of the popular view of the sexual instinct:

> It is generally understood to be absent in childhood, to set in at the time of puberty in connection with the process of coming to maturity and to be revealed in the manifestations of an irresistible attraction exercised by one sex upon the other; while its aim is presumed to be sexual union, or at all events actions leading in that direction ([11], p. 135).

But it quickly becomes obvious that this will not do as a definition of the sphere of the sexual. Sexuality is not confined to heterosexual genital intercourse between adults, for there are a number of perversions, and even popular opinion recognizes these as sexual in their nature. Popular opinion might wish to maintain a narrow conception of what is to count as *normal* sexuality, thus raising a problem about how one is to distinguish between normal and abnormal sexuality, but the more interesting and immediate problem is to make clear in virtue of what the perversions are recognized as sexual at all. And it is here that Freud makes an enormous conceptual advance. He distinguishes the object and the aim of the sexual instinct (decomposing what might have

E. E. Shelp (ed.), Sexuality and Medicine, Vol. I, 153–184.

seemed an indissoluble unity), and he introduces the notion of erotogenic zones (thus extending sexuality beyond the genitals), and is thus able to show that the perversions involve variations along a number of dimensions (source, object, and aim) of a single underlying instinct. Heterosexual genital intercourse is one constellation of variations, and homosexuality is another. Homosexuality, or inversion, involves variation in object, but the sexual sources (erotogenic zones, or bodily centers of arousal) and aims (acts, such as intercourse and looking, designed to achieve pleasure and satisfaction) may be the same. Thus what makes homosexuality recognizably sexual, despite its distance from what might be presented as the ordinary person's definition of sexuality, is the vast amount that it can be seen to have in common with 'normal' sexuality once one comes to understand the sexual instinct as itself complex, as having dimensions.

Freud makes the complexity of the sexual instinct compelling by drawing on the researches of the tireless investigators of sexual deviation such as Krafft-Ebing and Havelock Ellis. He makes the complexity intelligible by distinguishing the few dimensions (source, object, and aim) of the underlying instinct that are needed to lend order to the vast variety of phenomena, providing an illuminating new classificatory scheme. Once each of the perversions is understood as involving variation along one or more dimensions of a single underlying instinct, Freud is in a position to do at least two extraordinary things: First, to call into question the primacy of one constellation of variations over another. And second, to show that other phenomena that might not appear on the surface sexual (e.g., childhood thumbsucking) share essential characteristics with obviously sexual activity (e.g., infantile sensual sucking involves pleasurable stimulation of the same erotogenic zone, the mouth, stimulated in adult sexual activities such as kissing), and can be understood as being earlier stages in the development of the same underlying instinct that expresses itself in such various forms in adult sexuality. Freud is in a position to discover infantile sexuality. To briefly retrace the steps to this point: Perversions are regarded as sexual because they can be understood as variations of an underlying instinct along three dimensions (somatic source, object, and aim). The instinct has components, is complex or 'composite' ([11], p. 162). If adult perversions can be understood in terms of an underlying instinct with components that can be specified along several dimensions, then many of the activities of infancy can also be so understood, can be seen as earlier stages in the development of those components. But now I wish to focus on the newly problematic relation of normal and abnormal sexuality. Is one set of variations better or worse than another? The mere fact of difference, variation in content, is no longer enough once one cannot say one set of variations is somehow natural and others are not. Once one sees sexuality as involving a single underlying instinct, with room for variation along several dimensions, new criteria for pathology are needed. Moreover, insofar as variation is thought-dependent, rather than a matter of biological aberration, the question arises of whether there is such a thing as a pathology of sexual thought. Is there room for a morality of desire and phantasy alongside the ordinary morality governing action?

HOMOSEXUALITY

Freud initially distinguishes inversion from perversion. Inversion involves displacement of the sexual object from members of the opposite sex to members of the same sex. Inversion includes male homosexuality and lesbianism. Insofar as it involves variation in object only, it may appear less deviant than other sexual aberrations. But insofar as the point of singling out inversion is to contrast it with aberrations involving displacement in aim rather than object, it might as well include a wider range of aberrations, aberrations where displacement is to someone or something other than members of the same sex. From that point of view, bestiality, necrophilia, etc. are more like inversion than like the other aberrations – and Freud in fact treats them together as "deviations in respect of the sexual object" ([11], p. 136). If we include these less common and more troubling variations in object, inversion may no longer seem a less problematical form of sexual aberration. Moreover, the distinction between inversion and perversion tends to collapse as Freud discusses fetishism (is the deviation in object? in aim? – [11], p. 153). And it should be remembered that homosexuality is itself (like heterosexuality) internally complex, encompassing many different activities and attitudes. I shall use 'perversion' broadly, as Freud himself usually does, so that homosexuality counts as a perversion within Freud's classificatory scheme.

Is that a reproach? In the *Three Essays*, Freud states explicitly that it is inappropriate to use the word perversion as a 'term of reproach' ([11], p. 160). But that is in the special context of exploring the implications of his expanded conception of sexuality. In the case of Dora, published in the same year (1905) as the *Three Essays*, he refers to a phantasy of fellatio as "excessively repulsive and perverted" ([10], p. 52). A reproach seems built into the reference. It could be argued that Freud is forced to use the vocabulary of the view he wishes to overthrow, and that it carries its unwelcome connotations with it. Indeed, he in the same place argues that "We must learn to speak without indignation of what we call the sexual perversions – instances in which the sexual function has extended its limits in respect either to the part of the body concerned or to the sexual object chosen" ([10], p. 50). Perhaps Freud's own feelings, about the term if not the specific acts referred to, are ambivalent. The important question is what the appropriate attitude is and whether Freud's theory offers any light. So, again, let us consider homosexuality. Supposing it is a perversion, is that a reproach? Is the fact that it counts as a perversion a reason for disapproving of it in others or avoiding it oneself?

One could take the high ground and claim that it is pointless to disapprove what is not in a person's control, and then argue that choice of sexual object or sexual orientation is not in a person's control. But this does not really take one very far. Perhaps one has no or only marginal control over whether one contracts diabetes, but this does not stop us from recognizing that diabetes is a bad thing (while it does compel us to treat diabetes patients as victims). Even if we had an aetiological theory that assured us that homosexuality is not a matter of choice, and so perhaps not properly disapproved, that would not settle the question of whether it is a good or a bad thing (something we should avoid if we could). Moreover, even if sexual orientation is a given, outside the individual's control, what is given is a direction to desire. There remains the question of whether the individual should seek to control and suppress, or act on and express, the given desires.[1] Freud does not in fact take the high ground. His own aetiological views seem to leave open the extent of biological and other dispositional factors in leading to homosexuality. Whether homosexuality is innate or acquired is for him an open and a complex question ([11], p. 140). And, to whatever extent it is acquired, the conditions of its acquisition are also complex ([11], pp. 144f.). The so-called 'choice' of a sexual object is thus multiply obscure, and it is unclear to what extent the relevant causal conditions are within the individual's control (though one might also question whether and when control should be regarded as a condition of responsibility – see [43] and [36]). Freud nonetheless argues, on other grounds, that the 'perversity' of homosexuality gives no reason to condemn it:

> The uncertainty in regard to the boundaries of what is to be called normal sexual life, when we take different races and different epochs into account, should in itself be enough to cool the zealot's ardour. We surely ought not to forget that the perversion which is the most repellent to us, the sensual love of a man for a man, was not only tolerated by a people so far our superiors in cultivation as were the Greeks, but was actually entrusted by them with important social functions. The sexual life of each one of us extends to a slight degree – now in this direction, now in that – beyond the narrow lines imposed as the standard of normality. The perversions are neither bestial nor degenerate in the emotional sense of the word. They are a development of germs all of which are contained in the undifferentiated sexual disposition of the child, and which, by being suppressed or by being diverted to higher, asexual aims – by being 'sublimated' – are destined to provide the energy for a great number of our cultural achievements ([10], p. 50).

This passage actually contains at least two different types of argument. One is an appeal to universality across individuals, another an appeal to diversity across cultures. There is no doubt that sexual standards are culturally relative: different societies approve and disapprove of different sexual activities. But one might still wonder whether some societies are perverse in a pejorative sense. There is no avoiding direct consideration of the question of the criteria for perversion. Do they allow for something more than culturally relative, or even individually relative (whatever pleases one), judgments of sexual value?

CRITERIA OF PERVERSION

Once one accepts Freud's view of the complexity of the underlying sexual instinct, the old content criterion for perversion and pathology must be abandoned. As Freud writes, "In the sphere of sexual life we are brought up against peculiar and, indeed, insoluble difficulties as soon as we try to draw a sharp line to distinguish mere variations within the range of what is physiological from pathological symptoms" ([11], pp. 160–161).

It might seem simple enough to provide a sociological or statistical specification of perversion, but there are difficulties. For what precisely would the statistics reflect? One's questionnaires or surveys might seek to discover what the majority regards as perverse, but that would leave one wanting to know what perversion is (after all, members of the majority might in fact be applying very various standards). One might try to avoid direct circularity by, without mentioning the concept perversion, trying to elicit information revealing of which sexual desires the majority disapproves. But circularity re-emerges on this approach because there might be all sorts of different grounds for disapproval (aesthetic, moral, religious, political, biological, medical . . .), and what one wants is to single out those desires and practices which are disapproved of as (specifically) perverse. It appears one's questions and evidence would have already to be applying some standard of perversion in order to achieve that singling out. Parallel and further problems would apply to surveys of actual sexual practices. (Are perversions necessarily rare? If a practice became popular, would it therefore cease to be perverse? And if a practice were rare, e.g. celibacy or adultery, would that necessarily make it perverse?) Surely perversion is meant to mark only a certain kind of deviation from a norm. And there is another difficulty. For whatever method one uses, it will turn out that what counts as perversion will vary from society to society, will vary over time and place, in short, will be culturally relative. So insofar as one's concern is wider than the views of a particular society or group, insofar as it is a concern with general psychological theory, with the nature of human nature, no sociological approach will do. Moreover, insofar as one's concern is personal, or perhaps even therapeutic (unless one's standards of therapy are simply adaptation to local and contemporary prevailing norms), that is, if one is concerned to know how one ought to live one's life (including one's sexual life), a sociological approach will not do. For one's society may be wrong-headed, prejudiced, misguided, or in other ways mistaken. One has only one life to live. It might be necessary to resist one's society's demands or even to leave it. So one must look further.

Perhaps perversion can still be defined in terms of content if we are willing to start (again) with the popular view of normal sexuality as consisting of heterosexual genital intercourse between adults: then, any sexual desire or practice which goes beyond the body parts intended for sexual union, or that devotes too exclusive attention to a form of interaction normally passed through on the way to the final sexual aim, or which is directed at an object other than an adult member of the opposite sex, might be regarded as perverse.[2] One might insist on this stand independently of what the members of any particular society happen to think. But as we have seen, once one accepts Freud's analysis of the sexual in terms of a single, but complex, underlying instinct, while it becomes clear why the sexual perversions count as sexual, it becomes unclear why they are perverse. What privileges heterosexual genital intercourse between adults? Is there some further criterion that transcends individual societal views?

One might consider disgust. That is, we might try to pick out sexual activities to be condemned as perverse on the basis of a, presumably natural, reaction of disgust. So fellatio and cunnilingus might count as perverse because of disgust felt at oral-genital contact. Extensions of sexual activity beyond the genitals, alternative sources of sexual pleasure, would be perverse if disgust at them were sufficiently widespread. But disgust is itself generally culturally variable and often purely conventional. As Freud points out, "a man who will kiss a pretty girl's lips passionately, may perhaps be disgusted at the idea of using her toothbrush, though there are no grounds for supposing that his own oral cavity, for which he feels no disgust, is any cleaner than the girl's" ([11], pp. 151–152). Nonetheless, Freud seems to think that a content criterion can be preserved in certain extreme cases "as, for instance, in cases of licking excrement or of intercourse with dead bodies" ([11], p. 161). Perhaps some things, such as licking excrement, are thought to be objectively, universally disgusting. But perverse practices reveal that is not true, and Freud should know better.

Developmentally, children must learn to be disgusted at feces. This fact may not be obvious, but Freud was well aware of it. During the period of his earliest speculations about anal erotism, Freud wrote a fascinating letter to his friend Fliess:

> I wanted to ask you, in connection with excrement-eating . . . and animals, when disgust first appears in small children and whether there is a period in early infancy when no disgust is felt. Why do I not go to the nursery and – experiment? Because with twelve-and-a-half hours' work I have no time, and because the womenfolk do not back me in my investigations. The answer would be interesting theoretically ([29], p. 192, Letter 58 of February 8, 1897).

(This letter reminds us of how little Freud's theories about infantile sexuality were based on the direct observation of children. Which, to my mind, far from undermining his achievement – given its substantial confirmation by subsequent observations – makes it all the more remarkable.) The answer to his question was well known to Freud by the time he wrote the *Three Essays*. Children will play quite happily with their little turds, and as Freud writes, the contents of the bowels "are clearly treated as a part of the infant's own body and represent his first 'gift': by producing them he can express his active compliance with his environment and, by witholding them, his disobedience" ([11], p. 186). And Freud elsewhere develops the analogy between feces and other valued possessions, such as gold [12].[3] Disgust at the excremental is itself in need of explanation.

> Where the anus is concerned . . . it is disgust which stamps that sexual aim as a perversion. I hope, however, I shall not be accused of partisanship when I assert that people who try to account for this disgust by saying that the organ in question serves the function of excretion and comes in contact with excrement – a thing which is disgusting in itself – are not much more to the point than hysterical girls who account for their disgust at the male genital by saying that it serves to void urine ([11], p. 152).

It is true that Freud singles out disgust as one of the triumvirate of 'forces of repression' (disgust, shame and morality – [11], pp. 162, 178), and it may be that the forces of repression are ultimately instinctual and so present in every society, but that need not fix the content of the reaction. That is, it *may be* that everyone is necessarily (meaning biologically) bound to feel disgust at something, while still leaving room for variation in the objects of disgust. It should be no more surprising that the objects of disgust (as an instinct) are variable, than that the objects of sexual desire (as an instinct) are variable. So if the objects of sexual desire have no fixed or determinate content, neither do the objects of sexual disgust. We must look elsewhere if we are to find usable criteria for perversion and pathology.

Before looking elsewhere, we should note that there is another problem in a content criterion for perversion, which stems not from the variations we have been emphasizing, but from the universality we have mentioned only in passing. Freud points out that we can find apparently perverse desires not only in (otherwise admirable) other societies, but also within ourselves. In the case of homosexuality, he points out that our desires are responsive to external circumstances. Many will turn to homosexual pleasures given the appropriate favorable or inhibiting circumstances (e.g., "exclusive relations with persons of their own sex, comradeship in war, detention in prison . . ." – [11], p. 140). And even more strongly Freud concludes:

> Psycho-analytic research is most decidedly opposed to any attempt at separating off homosexuals from the rest of mankind as a group of a special character. By studying sexual excitations other than those that are manifestly displayed, it has found that all human beings are capable of making a homosexual object-choice and have in fact made one in their unconscious ([11], p. 145n.).

There is a sense in which all human beings are bisexual. Moreover, the universality of perversions other than homosexuality is exhibited in the role they play in foreplay ([11], pp. 210, 234). The prevalence of perversion (and the 'negative' of perversion, neurosis) receives its theoretical underpinning in terms of the universality of polymorphously perverse infantile sexuality. But for now the point is to see that a simple content criterion for perversion will not do. Given the facts of variety in cultural practice and of uniformity in individual potential, it is difficult to see how any particular object-choice (to focus on one dimension) can be singled out as necessarily abnormal. The nature of the sexual instinct itself sets no limit, for as Freud concludes, "the sexual instinct and the sexual object are merely soldered together" ([11], p. 148).

An alternative criterion for perversion and pathology emerges in connection with Freud's discussion of fetishism. Freud characterizes fetishism in general in terms of those cases "in which the normal sexual object is replaced by another which bears some relation to it, but is entirely unsuited to serve the normal sexual aim" ([11], p. 153). (Note that the variation seems to affect both object and aim.) But he shows that it has a point of contact with the normal through the sort of overvaluation of the sexual object, and of its aspects and of things associated with it, that seems quite generally characteristic of love. He continues:

> The situation only becomes pathological when the longing for the fetish passes beyond the point of being merely a necessary condition attached to the sexual object and actually *takes the place* of the normal aim, and, further, when the fetish becomes detached from a particular individual and becomes the *sole* sexual object. These are, indeed, the general conditions under which mere variations of the sexual instinct pass over into pathological aberrations ([11], p. 154).

Freud spells out the general conditions in terms of 'exclusiveness and fixation':

> In the majority of instances the pathological character in a perversion is found to lie not in the *content* of the new sexual aim but in its relation to the normal. If a perversion, instead of appearing merely *alongside* the normal sexual aim and object, and only when circumstances are unfavourable to *them* and favourable to *it* – if, instead of this, it ousts them completely and takes their place in *all* circumstances – if, in short, a perversion has the characteristics of exclusiveness and fixation – then we shall usually be justified in regarding it as a pathological symptom ([11], p. 161).

But this really will not do as a general criterion either, for reasons provided by Freud himself in a note a few pages earlier:

> psycho-analysis considers that a choice of an object independently of its sex – freedom to range equally over male and female objects – as it is found in childhood, in primitive states of society and early periods of history, is the original basis from which, as a result of restriction in one direction or the other, both the normal and the inverted types develop. Thus from the point of view of psycho-analysis the exclusive sexual interest felt by men for

women is also a problem that needs elucidating and is not a self-evident fact based upon an attraction that is ultimately of a chemical nature ([11], p. 146n.).

Once it is recognized that the instinct is merely soldered to its object, that there are wide possibilities of variation in the choice of object, then every choice of object becomes equally problematical, equally in need of explanation. Exclusiveness and fixation cannot be used to mark off homosexuality as perverse without marking off (excessively strong) commitments to heterosexuality as equally perverse. Thus, exclusiveness and fixation are no help if the point of a criterion for perversion is to distinguish the abnormal from the normal, and if heterosexual genital intercourse between adults is to be somehow privileged as the paradigm of the normal. We need some norm for sexuality if the notion of perversion is to take hold. From where can we get it? Is there any reason to suppose that it will take the form of the popular view of normal sexuality?

DEVELOPMENT AND MATURATION

Freud in fact, as we have seen, operates with multiple criteria for perversion and pathology. We have also seen that his own views provide materials for a critique of those criteria if one attempts to generalize them. But there emerges from within his theory yet another criterion, a criterion which is meant to be ultimately biological and so not culturally relative. As Freud puts it at the start of the third of his *Three Essays*: "Every pathological disorder of sexual life is rightly to be regarded as an inhibition in development" ([11], p. 208). Perverse sexuality is, ultimately, infantile sexuality. While consideration of the adult perversions prepares the way for the extension of our understanding of sexuality to infantile activities in the course of Freud's book, infantile sexuality prepares the way for both normal and perverse sexuality in the development of the individual.[4] It is through arrests in that development, or through regression to earlier points of fixation when faced by later frustration, that an adult comes to manifest perverse sexual activity. We can pick out sexual desires and activities which count as perverse if we have an ideal of normal development and maturation.

Freud's theory of psychosexual development, with its central oral-anal-genital stages, provides such an ideal. The dynamic is at least partly biological. At first, the infant has control of little other than its mouth, and in connection with its original need for taking nourishment it readily develops independent satisfaction in sensual sucking ([11], p. 182). That the anus in due course becomes the center of sexual pleasure and wider concerns ("holding back and letting go") is not surprising in the light of a variety of biological developments: as the infant gets older, the feces are better formed, there is more sphincter control (so the child begins to have a choice about when and where to hold back or let go), and with teething there is pressure for the mother to wean.[5] Finally, there comes puberty and the possibility of reproduction and increased interest in the genitals. But one should not totally biologize what is at least in part a social process. There may be a confusion between the ripening of an organic capacity with the valuation of one form of sexuality as its highest or only acceptable form. The subordination of sexuality to reproduction, and the importance attached to heterosexual genital activity, is after all, a social norm. Freud does not claim that there is a biological or evolutionary *preference* for reproduction; the individual preference, if any, is simply for end-pleasure. Even if the preference for end-pleasure or orgasm over fore-pleasure ([11], pp. 210–212) is biologically determined, the conditions for such pleasure are not. Whether end-pleasure takes place under conditions that might lead to reproduction depends on a wide range of factors, and whether it *should* take place under such conditions is subject to both circumstance and argument. Even if one attaches supreme importance to the survival of the species, other things, including sexual pleasure (which may in turn depend on a certain degree of variety) may be necessary to the survival of the species. And for most of recent history, over-population and unwanted conception have been of greater concern than maximizing the reproductive effects of sexual activity. Under certain circumstances homosexuality might have social advantages.[6]

In terms of Freud's instinct theory (not to be confused with standard biological notions of hereditary behavior patterns in animals), every instinct involves an internal, continuously flowing source of energy or tension or pressure. Freud adds, however: "Although instincts are wholly determined by their origin in a somatic source, in mental life we know them only by their aims" ([21], p. 123). Given Freud's fundamental hypotheses concerning the mechanisms of psychic functioning, the aim is in every case ultimately discharge of the energy or tension. And given Freud's discharge theory of pleasure (or tension theory of unpleasure), the aim must ultimately be understood in terms of pleasure. Freud is well aware of the problems of a simple discharge theory of pleasure, especially in relation to sexuality (where, after all, the subjective experience of increasing tension is typically as pleasurable as the experience of discharge). (See [11], pp. 209f., and [26].) The point here, however, is that on Freud's view the essential aim of sexual activity (as instinctual activity) must be pleasure, achievable by a wide variety of particular acts (under a wider variety of thought-dependent conditions). Sexuality may serve many other purposes and have many other functions and aims from a range of different points of view. Among these are reproduction, multi-level interpersonal awareness, interpersonal communication, bodily contact, love, money. . . .[7] Within Freud's theory, perversion is to be understood in terms of infantile, that is non-genital, forms of pleasure. This approach has its problems. For one thing, homosexuality, in some ways the paradigm of perversion for Freud, is not necessarily non-genital and so not obviously perverse by this criterion. Moreover, insofar as other perversions, such as fetishism, aim at genital stimulation and discharge, they too are not purely infantile. (Cf. [22], p. 321.) In practice, of course, Freud collapses the individual's experienced concern for genital pleasure together with the biological function of reproduction, so that the development and maturation criterion for perversion reduces to the question of the suitability of a particular activity for reproduction.

One should not confuse the (or a) biological function of sexuality, namely reproduction, with sexuality as such. Freud is at pains to point out that sexuality has a history in the development of the individual that precedes the possibility of reproduction. The reproductive function emerges at puberty ([22], p. 311). An ideal of maturation that gives a central role to that function makes all earlier sexuality of necessity perverse. The infant's multiple sources of sexual pleasure make it polymorphously perverse. And the connection works both ways. Sexual perversions can be regarded as in their nature infantile. As Freud puts it:

> if a child has a sexual life at all it is bound to be of a perverse kind; for, except for a few obscure hints, children are without what makes sexuality into the reproductive function. On the other hand, the abandonment of the reproductive function is the common feature of all perversions. We actually describe a sexual activity as perverse if it has given up the aim of reproduction and pursues the attainment of pleasure as an aim independent of it.

> So . . . the breach and turning-point in the development of sexual life lies in its becoming subordinate to the purposes of reproduction. Everything that happens before this turn of events and equally everything that disregards it and that aims solely at obtaining pleasure is given the uncomplimentary name of 'perverse' and as such is proscribed ([22], p. 316).

I believe Freud may well provide an accurate account of the link in our language between perversion and non-reproductive sex. On the other hand, I don't believe Freud's theory is committed to maintaining that link (the theoretically necessary aim is pleasure, not reproduction). Moreover, even if detachment from the possibility of reproduction is a necessary condition of regarding a practice as perverse, it cannot be sufficient: otherwise sterile heterosexual couples or those who use contraceptives would have to be regarded as perverse. (More on these matters in a moment.)

In privileging heterosexual genital intercourse between adults, if only for the purpose of classifying the perversions, one is making a choice based on norms. Freud's discussion of reproduction reflected existing social norms, and so the fact that they were norms was perhaps concealed. The norms of the sexual liberationists, such as Herbert Marcuse and Norman O. Brown, are in some ways perhaps continuous with the standards built into Freud's model. Does polymorphous perversion include sadism? Should it? Contemporary debates over the appropriate ideals of sexuality cannot be decided by simple appeals to biology. 'Regression' is doubtless an empirical concept, but it gets its sense against a background provided by social norms of development (not purely biological norms of development). In picking out the perversions we apply an external standard to sexuality. Which is not to say that we should not. It is to say only that we should be self-conscious about what we are doing and why. Calling perversions 'infantile' may in fact describe them, but the immature is usually regarded as inferior. And if that judgment is to follow, one needs more grounds than those provided by biology. After all, if we live long enough, we eventually decay. Later does not necessarily mean better.

MORE ON HOMOSEXUALITY

Is homosexuality a perversion? On a content criterion, whether ultimately based on a reaction of disgust or something else, the answer will vary over time and place, and it is arguable that the reaction of disgust is at least as malleable as the desire to which it is a reaction. On a criterion of exclusiveness and fixation, it is no more or less a perversion than heterosexuality of equivalent exclusivity. On a criterion of development and maturation, or arrest and regression, the answer is less clear. Many say that homosexuality is a developmentally immature stage or phase. I do not see where Freud says that. In the *Three Essays*, Freud notes that homosexuality "may either persist throughout life, or it may go into temporary abeyance, or again it may constitute an episode on the way to a normal development." He goes on, "It may even make its first appearance late in life after a long period of normal sexual activity" ([11], p. 137). In this case, it is heterosexuality that is the earlier phase. In passing, in the lecture on anxiety in the *New Introductory Lectures on Psycho-Analysis*, Freud indicates that "in the life of homosexuals, who have failed to accomplish some part of normal sexual development, the vagina is" represented by the anus ([28], p. 101) and presumably therefore avoided. But this must refer to only one type of homosexual (and surely not the kind that prefers sodomy).[8] Freud does say that infantile sex is characteristically auto-erotic ([11], p. 182), that is, involves no sexual object. In that respect, homosexuality is clearly not infantile. But then foot fetishism and bestiality also involve objects. Would one want to conclude that they are also not infantile, also not perverse? The presence of a whole person as object in the case of homosexuality doubtless makes a significant difference. (Inversion as such may, after all, be importantly different from perversion as such.)

The closest Freud comes to referring to homosexuality as an immature form of sexuality is in a letter in response to a mother who wrote him about her homosexual son. Freud wrote:

> Homosexuality is assuredly no advantage, but it is nothing to be ashamed of, no vice, no degradation; it cannot be classified as an illness; we consider it to be a variation of the sexual function, produced by a certain arrest of sexual development. Many highly respectable individuals of ancient and modern times have been homosexuals, several of the greatest men among them (Plato, Michelangelo, Leonardo da Vinci, etc.). It is a great injustice to persecute homosexuality as a crime – and a cruelty, too. . . . What analysis can do for your son runs in a different line. If he is unhappy, neurotic, torn by conflicts, inhibited in his social life, analysis may bring him harmony, peace of mind, full efficiency, whether he remains homosexual or gets changed ([7], pp. 419–420, April 4, 1935).

Without support from his theoretical writings, the "arrest of sexual development" must be presumed to refer to (the social norm of) reproduction. At a theoretical level, it is only in the case of lesbianism that there looks like there is a stage-specific point to be made about object-choice. That is, given the basic premises of psychoanalytic theory, it is not entirely clear why all women are not lesbians. Up to the genital phase, their development parallels that of little boys, and the beginnings of object relations should tie both little boys and girls to their mothers as the main supporting figure. Girls, unlike boys, are supposed to switch the gender of their love objects in the course of going through their Oedipal phase. The incest taboo is supposed to lead boys to exclude their mothers, but not all women, as possible sexual objects. Under pressure of the castration complex, and through identification with their father, boys are supposed to search for "a girl just like the girl who married dear old dad". Girls, on the other hand, are supposed to switch from a female to a male love object. Why they do this is open to various accounts: Some in terms of penis envy (which needs more elaboration than can be provided here – in any case, biological accounts in terms of a switch in interest from clitoris to vagina will not work). Some in terms of rivalry with the same-gender parent (something girls have in common with boys – it is just that their same-gender parent happened previously to have been the primary object of dependence and so love). Some in terms of a desire to please the mother (involving getting a penis for her). Whatever the account one gives of female psychosexual development, there is little reason to regard male homosexuality as involving arrest at or regression to an earlier phase of development, and so as infantile and (on that criterion) perverse.[9]

Still, perhaps something further can be extracted from Freud's general theory of development. It might be argued that there is a sense in which the basic mechanism of homosexual object-choice is more primitive than the mechanism involved in heterosexual choice. Freud distinguishes two basic types of object-choice: anaclitic and narcissistic ([20], pp. 87–88). On the anaclitic (or attachment) model, just as the sexual component instincts are at the outset attached to the satisfaction of the ego-instincts, the child's dependence on the parents provides the model for later relationships. On the narcissistic model, the individual chooses an object like himself. It might seem obvious that homosexual object-choice is narcissistic, and that narcissistic object-choice is more primitive than the other type. Neither point is correct. While the homosexual certainly has an object that is in at least one respect (gender or genitals) like himself, there are many other aspects of the individual, and in terms of those other aspects even heterosexual object-choice can be importantly narcissistic. Moreover, the mechanisms of homosexual object-choice are various (e.g., Freud sometimes gives emphasis to the avoidance of rivalry with the father or brothers), and the similarity of the object to oneself may not be crucial in all cases – indeed, an anaclitic-type dependence on the object may be much more prominent.[10] That narcissism as a stage, in the sense of taking oneself as a sexual object, may be more primitive than object-choice, in the sense of taking someone else as a sexual object, does not make the narcissistic type of object-choice more primitive than the anaclitic type. In both cases, unlike primitive narcissism, someone else is the object, it is just that on one model similarity matters most, on the other dependence matters most. Even if narcissism is considered the first form of object-choice (after auto-erotism), dependence is present from the very beginning (and a whole school of psychoanalysis would argue object relations are present from the very beginning). Freud himself wrote:

> At a time at which the first beginnings of sexual satisfaction are still linked with the taking of nourishment, the sexual instinct has a sexual object outside the infant's own body in the shape of his mother's breast. It is only later that the instinct loses that object, just at the time, perhaps, when the child is able to form a total idea of the person to whom the organ that is giving him satisfaction belongs. As a rule the sexual instinct then becomes auto-erotic, and not until the period of latency has been passed through is the original relation restored. There are thus good reasons why a child sucking at his mother's breast has become the prototype of every relation of love. The finding of an object is in fact the refinding of it ([11], p. 222).

Homosexuality is no *more* a return to earlier modes of relationship than any other attempt at love.[11]

The American Psychiatric Association has struggled with the question of the classification of homosexuality. The classification is not without practical implications, and it is not surprising that the debate has taken political turns.[12] Nosology is not simply a matter of aetiological theories in any case. At the minimum, classification sometimes takes account of symptomatic patterns and treatment possibilities as well as aetiology. The argument against classifying homosexuality as a disease could well include the notion that it *should not* be treated (whatever its origin) as well as the political claim that the disease classification contributes to inappropriate discrimination (e.g., in jobs – should homosexuality be grounds for dismissal? should schizophrenia?). In 1973, the Board of Trustees of the American Psychiatric Association voted to remove homosexuality (as such) from the list of disorders in the *Diagnostic and Statistical Manual of Mental Disorders* ([5], pp. 281–282). Nonetheless, something called 'ego-dystonic homosexuality' was included. That is, if a homosexual does not desire his condition, or suffers distress at his condition, the condition is then regarded as a disorder. Clearly the criteria of mental disorder employed by the APA in this connection are not 'neutral': distress and undesirability can be traced to social attitudes (what produces distress and is therefore undesired in Iowa may be very different from what produces distress and is undesired in San Francisco – so homosexuality might be a 'disorder' in Iowa but not San Francisco).[13] In any case, it does not follow from the aetiological and developmental theories of psychoanalysis that homosexuality must produce distress and so be undesired.

It must be acknowledged, however, that even if homosexuality involves no developmental arrest or inhibition, even if homosexuality is as 'genital' and mature as heterosexuality, it is, as things currently are, detached from the possibility of reproduction and in *that* sense perverse. Any sexual activity which must be detached in its effect from reproduction can be, and has been, regarded as perverse. (Note the relevant detachment is in effect, not in purpose. If the purpose of the persons engaged in the activity was what mattered, most heterosexual genital intercourse would have to be regarded as perverse.) Granting this sense to perversion, however, one should be careful what one concludes about people whose activities are in this sense perverse. For one thing, reproduction would in fact be excluded only if their activities were exclusively perverse. For another, whether it is socially beneficial to *bear* children (the care and upbringing of children is not excluded by perverse – that is, non-reproductive – activity) depends on circumstances (other features of the parents, and social circumstances such as over-population). Moreover, new reproductive technologies may make the reproductive limitations of perverse activity of lesser concern, just as new contraceptive technologies have made the dangers of unwanted conception of lesser concern in 'normal' sexual activity. Whatever the biological place of reproduction in human sexual life, it cannot settle the appropriate attitude to non-reproductive human sexual activity. After all, normal sex, that is, heterosexual genital intercourse between adults, can be multiply defective. There can be failures of reciprocity and mutuality, or interactive completeness (private sexual phantasies may make intercourse closer to masturbation in its experience, even if not in its possible effects). And even sex normal in the present sense, that is, of the kind that could in appropriate circumstances lead to reproduction, may fail in its actual effects (most intercourse does not lead to pregnancy, and intercourse between sterile partners or involving the use of contraceptives is most unlikely to). Does detachment from reproductive concerns in one's sexual activity make an individual defective? There is no reason to believe so. Freud frequently points out the great social contributions of homosexuals in history, sometimes even tying the contributions to the sexual orientation, deriving social energies from homosexual inclinations.[14] Not that Freud is blind to defects; he does not assume all homosexuals are mainstays of civilization: "Of course they are not . . . an 'elite' of mankind; there are at least as many inferior and useless individuals among them as there are among those of a different sexual kind" ([22], p. 305). Whether homosexuals contribute to society may be relevant to the question of the appropriate attitude to take towards homosexuality, but the same can be said for heterosexuals and those of mixed inclinations; there is no reason to expect uniformity of contributions within such groupings. It remains unclear whether homosexuality should be regarded as a perversion: it depends on which criterion for perversion is adopted (e.g. content, with disgust the marker; exclusiveness and fixation; or development and maturation, with reproduction the marker), and given certain criteria, on which developmental and aetiological theories are believed. But it does seem clear that even if homosexuality is regarded as a perversion, that in itself gives no ground for condemning it or thinking it worse than heterosexuality; no reason to disapprove it in others or avoid it in oneself.

FOOT FETISHISM

If anything is a perversion according to prevailing attitudes, foot fetishism is, and Freud's discussion of exclusiveness and fixation helps us understand why.[15] But other criteria of perversion (content, maturation, reproduction, completeness . . .) would doubtless yield the same result – indeed, it might be a condition of adequacy on such criteria that they yield that result. Classification is not the problem. Understanding the source and point of this sort of unusual interest in feet is.

Usually, when confronted with a desire one does not share, one can sympathize with the unshared desire at least to the extent of having a sense of what is desirable about the object. Part of the mystery of fetishism is making sense of the extraordinary value and importance attached to the object. Bringing out the link of fetishism to more ordinary overvaluation of sexual objects (which can in turn be tied to narcissism – [20], pp. 88–89, 91, 94, 100–101) goes some way towards making fetishism intelligible ([11], pp. 153–154), but it still leaves us wanting to know why desires should take such peculiar directions. Partly this is a question about the mechanism of object-choice, but more importantly it is a question about the meaning of object-choice. What is it about a foot that makes it so attractive? Why are some particular feet more attractive than others? How can they come to satisfy (or be seen to satisfy) needs? Psychoanalysis offers answers. In the central cases, "the replacement of the object by a fetish is determined by a symbolic connection of thought, of which the person concerned is usually not conscious" ([11], p. 155). In the case of foot fetishism, in condensed form, psychoanalysis argues (among other things) that "the foot represents a woman's penis, the absence of which is deeply felt" ([11), p. 155n.). Thus condensed the answer may seem wildly implausible. But in his paper on fetishism [27] Freud traces a chain of experience, phantasy, and association, that suggests how a foot might come to provide reassurance about castration fears, and so become the focus for sexual interests. Thus filled in, the story may still seem implausible. But notice that the question of plausibility enters at two levels: one is the plausibility of the beliefs ascribed to the fetishist (how could anyone believe anything as implausible as that a foot is the mother's missing penis?), and the second is the plausibility of the ascription of the (implausible) beliefs. The genius of the psychoanalytic account is not that it seeks to make bizarre or ad hoc beliefs plausible, but it takes beliefs that it gives us other reasons for ascribing to people and shows how in certain cases they persist and give direction to desire.

Some of the relevant beliefs (e.g., in the ubiquity of the male genital) are to be found in infantile sexual theories. Much of the evidence for such beliefs, as well as for symbolic equations, comes from the study of neurotics; which is as it should be, for, as Freud repeatedly points out, "neuroses are . . . the negative of perversions" ([11], p. 165). We should perhaps pause for a moment on this point. The sexual instinct, we have seen, is complex, has several dimensions ([11], p. 162). It is not the simple, 'qualityless' energy of much of Freud's earliest theorizing ([11], pp. 168, 217). It is thus possible to reidentify the 'same' instinct in different contexts because variation in (for example) object may leave the source clearly the same. Instincts, unlike qualityless energy, meet one of the conceptual restrictions on 'displacement': a change in object can be seen as 'displacement' (rather than mere change) only against a background of continuity. One of the things that may have concealed the underlying continuity between infantile and adult sexuality is that the infant is 'polymorphously perverse' ([11], p. 191) – and the tie to adult sexuality is clearest in relation to perverse sexuality (not heterosexual genital intercourse). Similarly, the role of sexuality in the neuroses was concealed partly because the sexuality involved is typically perverse: as Freud puts it, "*neuroses are, so to say, the negative of perversions*" ([11], p. 165) – so the sexual nature of neuroses tends to be hidden. What Freud means by the famous formula is spelled out a bit more fully in a note: "The contents of the clearly conscious phantasies of perverts (which in favourable circumstances can be transformed into manifest behaviour), of the delusional fears of paranoics (which are projected in a hostile sense on to other people) and of the unconscious phantasies of hysterics (which psycho-analysis reveals behind their symptoms) – all of these coincide with one another even down to their details" ([11], p. 165 N. 2). To make this claim persuasive, one must bring out the content of the unconscious phantasies of hysterics, but this is made simpler by the fact that, in the case of neurotics, "the symptoms constitute the sexual activity of the patient" ([11], p. 163), and "at least *one* of the meanings of a symptom is the representation of a sexual phantasy" ([10], p. 47). Thus Dora's hysterical cough could be analyzed in terms of an unconscious phantasy of fellatio ([10], pp. 47–52). None of this is very surprising if one remembers that neurotic sexuality, like perverse sexuality, is infantile ([11], p. 172) – whatever shape the sexual instinct eventually takes, it inevitably has its roots in infantile sexuality.

Returning to foot fetishism, whatever one thinks of the psychoanalytic story, it is clear that some story is needed. The attachment is, without further explanation, too peculiar. It is hard for one who does not share the desire to see what is desirable. With suitable hidden significances, the desire at least becomes intelligible as desire. And to understand all may here be to forgive all, if forgiveness is needed. By the standard of exclusiveness and fixation, fetishism is doubtless perverse. We have argued that the criterion of exclusiveness and fixation is itself inadequate if applied quite generally. Nonetheless, there is something peculiar about fetishism, and insofar as psychoanalysis can help us understand that peculiarity, it may help us understand the appropriate attitude towards perversions in general. In the case of fetishism, while we might not share the beliefs, we can see how given certain beliefs, certain objects and activities might become desirable. It does not follow that all desires become equally uncriticizable once understood. The beliefs may have wider implications and having the beliefs and desires may have wider effects. So some perversions may be objectionable. Our ordinary standards for judging human action and human interaction do not lapse in the face of perversions; but the mere fact of perversion is not an independent ground for moral criticism.

Again, foot fetishism demands some explanation. Those who wish to reject the psychoanalytic account of foot fetishism have the burden of supplying an alternative. I believe that a simple stimulus generalization account will not do. Psychoanalysis readily includes the standard associationist points, though sometimes adding less standard associative connections as well; for example, Freud notes:

> In a number of cases of foot-fetishism it has been possible to show that the scopophilic instinct, seeking to reach its object (originally the genitals) from underneath, was brought to a halt in its pathway by prohibition and repression. For that reason it became attached to a fetish in the form of a foot or shoe, the female genitals (in accordance with the expectations of childhood) being imagined as male ones ([11], p. 155 N. 2; cf. [27] p. 155).

But Freud is also properly wary of attributing too much to early sexual impressions, as though they were the total determinant of the direction of sexuality:

> All the observations dealing with this point have recorded a first meeting with the fetish at which it already aroused sexual interest without there being anything in the accompanying circumstances to explain the fact. . . . The true explanation is that behind the first recollection of the fetish's appearance there lies a submerged and forgotten phase of sexual development. The fetish, like a 'screen-memory', represents this phase and is thus a remnant and precipitate of it ([11], p. 154 N. 2).[16]

The connections Freud emphasizes are typically meaningful, rather than mere casual associations. The more general problem with simple stimulus generalization is that it tends to explain both too little and too much. Why do other people exposed to the same stimuli not develop fetishistic attachments? (Psychoanalysis may also have trouble with this question. See [27], p. 154.) Why do fetishists often attach special conditions (such as smell) to their preferred objects? (Here psychoanalysis has some interesting suggestions. See [15], p. 247; and [11], p. 155 N. 2.) If stimulus generalization stands alone as an explanatory mechanism, it can appear able to explain actual particular outcomes of an association only at the expense of appearing equally able to explain any other outcome of a given early impression. The factors pointed to by the conditioning theorists are simply too pervasive and nondiscriminating. Something that would explain everything explains nothing. (See [37], pp. 126–127.)

The desires of the fetishist are typically highly thought-dependent. He sees the fetish object as of a certain kind, as having certain connections. (This 'seeing as' is another aspect of the situation generally neglected by behaviorist approaches. See [42].) Psychoanalysis seeks to trace out these connections (some of them hidden from the individual himself) and their history. It seeks to understand their compulsive force and to enable the individual to specify more fully what it is that he desires in relation to the object. The thought of the object (including the thought of the reason for the desire or of the feature that makes the object desired desirable) specifies the desire. A proper understanding of the relevant thoughts may be a necessary condition of freedom, of the possibility of altering desire via reflective self-understanding. A too exclusive attention to the behavior involved in perverse sexuality may neglect the thought and so the desire behind the behavior. Since people may do observably the same thing for very different reasons (sometimes one person wants to, while another person might be paid to; the different meanings of the same behavior may be revealed in associated phantasies, conscious and unconscious, and other thoughts), behaviorist specifications of perverse activity, like sociological accounts of perverse activity, may inevitably miss the point. If we are to understand perverse (and also 'normal') sexual desires (and activities) we must look to the thoughts behind them.[17]

THE MENTAL AND THE PHYSICAL

Plato draws a line between physical love and spiritual love, thinking the latter higher than the former.

The line between the physical and the mental does not correspond to the line between the sexual and the spiritual. For whatever one thinks of spirituality and mentality, sexuality is not purely physical. Indeed, if it were, one might expect the objects and aims of sexual desire to be fixed by biology. But while human biology is relatively uniform, the objects and aims of sexual desire are as various as the human imagination. There are psychological conditions of sexual satisfaction. Sex is as much a matter of thought as of action. While the machinery of reproduction, the sexual organs themselves, the genitals, have determinate structures and modes of functioning, sexual desire takes wildly multifarious forms. Sexuality is as much a matter of thought or the mind as of the body. To think one can get away from sexuality via the denial of the body is to mistake the half for the whole.

While it would be an exaggeration to say sex is all in the mind, it would be less of a mistake than the common notion that sex is purely physical. Freud came closest to the truth in locating sexuality at the borderland or bridge between the mental and the physical. Writing of instincts in general, Freud explained his meaning:

> By an 'instinct' is provisionally to be understood the psychical representative of an endosomatic, continuously flowing source of stimulation, as contrasted with a 'stimulus', which is set up by *single* excitations coming from *without*. The concept of instinct is thus one of those lying on the frontier between the mental and the physical ([11], p. 168).

Thus the sexual instinct is not to be equated with neutral energy (as in Freud's earlier theorizing, e.g. in [8]). It has direction (aim and object) as well as a somatic source and impetus (or strength). The instinct involves both biologically given needs and thought-dependent desires. It is our thoughts that specify the objects of our desire (however mistaken we may be about whether they will satisfy our real needs). Via transformations and displacements of various sorts, our sexual instinct takes various directions. As Freud at one place puts it, "In psycho-analysis the concept of what is sexual . . . goes lower and also higher than its popular sense. This extension is justified genetically . . ." ([17], pp. 222; cf. the discussion of 'The Mental Factor' at [11], pp. 161–162). The analysis of sexual desires starts with an instinctual need derived from a somatic source. But the psychical representatives of this instinctual need develop in the history of the individual, attracting him to a variety of objects and aims (modes of satisfaction). Given different vicissitudes, our original instinctual endowment develops into neurosis, perversion, or the range of normal sexual life and character. Our character is among those (perhaps 'higher') attributes that Freud traces back to sexuality. In his essay on 'Character and Anal Erotism' Freud says we can "lay down a formula for the way in which character in its final shape is formed out of the constituent instincts: the permanent character-traits are either unchanged prolongations of the original instincts, or sublimations of those instincts, or reaction-formations against them" ([12], p. 175). I cannot pursue the puzzles raised by these alleged transformations, and by the psychoanalytic explanation of the normal, here (I make a start in [37], esp. pp. 191–192), but it should be clear that our sexual character in large measure determines our character, who we are: whether directly, as suggested in the formula, or indirectly, as the model for our behavior and attitudes in other spheres.[18]

There are lessons in multiplicity to be learned from Freud. At a minimum, I would have us take the following from this essay on Freud's *Three Essays*:

First. Sexuality, far from being unified, is complex. The sexual instinct is made up of components which can be specified along several dimensions (source, object, aim). It is a composite that develops and changes, and can readily decompose. In particular, the instinct is 'merely soldered' to its object.

Second. The criteria for perversion are multiple, and no one of them is truly satisfactory if one is searching for a cross-cultural standard founded in a common human nature. Not that there are not ideals of sexuality (with corresponding criteria for perversion), but they too are multiple, and must be understood in connection with more general ideals for human interaction.

Third. The purposes, functions, and goals of sexuality are multiple. It is not a pure bodily or biological function. There is a significant mental element that emerges perhaps most clearly in relation to the perversions, where the psychological conditions for sexual satisfaction are dramatically emphasized. Here we might find the beginnings of a defensible (Spinozist-Freudian) ideal in the sphere of the sexual: health and maturity involve coming to know what we really want and why we want it. Further, since what we want depends on what we think, if we wish to change what we want, we may have to change how we think.

Who we are is revealed in who or what and how we love. The structure of our desires emerges in the course of the transformation of the sexual instinct as we learn to live in a world full of internal and external pressures and constraints, as we learn to live with others and ourselves.

University of California,
Santa Cruz, California,
U.S.A.

NOTES

[1] While I here emphasize that the existence of some causal story does not render all evaluation out of place, I should perhaps also emphasize that some evaluations are almost always out of place. Whether homosexuality is the result of nature or nurture, it makes little sense to condemn homosexuality as 'unnatural'. For one thing, nature, or at least human nature, includes conditions of nurture: all humans must be somehow nurtured in order to survive and develop. The 'somehow' of course allows for variations. The real point of the contrast of nature and nurture, two types of causes, may ultimately simply be in terms of uniformity versus variability. In terms of individual responsibility, nature and nurture may both be viewed as 'external' causes (the individual does not choose them, and so does not control the result). For another thing, nature in general includes more than many would like to admit (one of the constant lessons of the Marquis de Sade). Insofar as charges of perversion are based on notions of unnaturalness, they may always be inapplicable. (See Michael Slote, 'Inapplicable Concepts and Sexual Perversion' [40].) The various contrasts between the natural and the unnatural, and the historical development of the charge of unnaturalness against homosexuality, are interestingly traced by John Boswell in his *Christianity, Social Tolerance, and Homosexuality* [4]. In the coroner's verdict, "death by natural causes", the contrast is with other types of causes, basically causes involving the intervention of human intentions. Whatever the causes of homosexuality and homosexual desires they must be of the same *type* as the causes of heterosexuality and heterosexual desires. This point is reflected in Aristophanes' myth in Plato's *Symposium*. Incidentally, one might note that if Freud had this myth in mind in his discussion at the start of the *Three Essays* ([11], p. 136), his account there is somewhat misleading. Freud speaks as if the 'poetic fable' is supposed to explain only heterosexuality, and as if the existence of homosexuality and lesbianism therefore comes as a surprise. In fact, Aristophanes' story of the division of the original human beings into two halves, and their subsequent quest to reunite in love, allows for all three alternatives: Aristophanes starts with three original sexes. Thus the myth offers an explanation (the same explanation) of homosexuality and lesbianism as well as heterosexuality. (One should perhaps also note that there is an Indian version of the myth that may conform better to Freud's account, and Freud refers to it explicitly later in *Beyond the Pleasure Principle*.) From the point of view of psychoanalytic theory, heterosexual object-choice and homosexual object-choice are equally problematic, equally in need of explanation ([11], p. 146n.).

Freud himself, in his published writings, only used the term 'unnatural' three times in connection with perverse desires or practices. In each of the three instances ([9], p. 265; [22], p. 302; and [23], p. 149), in context, the term refers to the views of others.

[2] Freud spells out the content criterion for deviations in respect of source and aim: "Perversions are sexual activities which either (a) extend, in an anatomical sense, beyond the regions of the body that are designed for sexual union, or (b) linger over the intermediate relations to the sexual object which should normally be traversed rapidly on the path towards the final sexual aim" ([11], p. 150). The question remains, what is so objectionable about 'extending' and 'lingering'?

[3] Freud summarizes his views on the child and feces in Introductory Lecture XX: "To begin with . . . He feels no disgust at his faeces, values them as a portion of his own

body with which he will not readily part, and makes use of them as his first 'gift', to distinguish people whom he values especially highly. Even after education has succeeded in its aim of making these inclinations alien to him, he carries on his high valuation of faeces in his estimate of 'gifts' and 'money'. On the other hand he seems to regard his achievements in urinating with peculiar pride ([22], p. 315).

[4] "Not only the deviations from normal sexual life but its normal form as well are determined by the infantile manifestations of sexuality" ([11], p. 212).

[5] Hence, as Erikson suggests, the infant is expelled from the oral paradise of an earlier stage ([6], p. 79). Erikson is in general very helpful on the social contribution to and meaning of the psychosexual stages.

[6] There has been some speculation on the possible evolutionary advantages of homosexuality in terms of altruistic and social impulses. (See, e.g., [44], pp. 142f.)

[7] The multiplicity of ends and essences for sexuality, and the corresponding multiplicity of criteria for perversion, is amply evidenced in a growing philosophical literature on sexual perversion (much of it collected in two anthologies: [1] and [41]). The authors tend to vacillate between on the one hand explicating the concept of perversion in a way which captures our ordinary classifications of particular practices, and on the other providing a sustained rationale for a defensible ideal of sexuality (with its attendant, sometimes revisionary, implications for what counts as a perversion). Here, as elsewhere, a 'reflective equilibrium' between our intuitions and principles may be desirable. Perhaps most interesting from the point of view of the issues considered in this essay are Thomas Nagel's 'Sexual Perversion' [35] and Sara Ruddick's 'Better Sex' [39]. Nagel finds the essence of sexuality in multi-leveled personal interaction and awareness, a dialectic of desire and embodiment that makes desires in response to desires central to sexuality. Hence the criterion for perversion that emerges is in terms of interactive incompleteness – according to which homosexuality need not be perverse, foot fetishism must be, and heterosexual intercourse with personal phantasies might be. While the form of incompleteness is different, the emphasis on incompleteness might be suggestively connected with the sort of unification or totalization of components in Freud's final genital organization of sexuality – in terms of which perversions might be understood as component (or 'incomplete') instincts. (Cf. Freud's statement, echoed often elsewhere, that the perversions are "on the one hand inhibitions, and on the other hand dissociations, of normal development" [11], p. 231.) In any case, Nagel's emphasis on a full theory of the nature of sexual desire seems to me right-headed. Also of special interest is Ruddick's 'Better Sex', which, among other things, sorts out clearly the relation of reproduction to perversion in ordinary language and understanding.

Freud's emphasis on the role of pleasure (or discharge) in sexuality should be complicated by his emphasis on the psychological conditions of pleasure (thought-dependent conditions of discharge). Pleasure, as Freud well understood, is not itself simply bodily or otherwise simple. When the question shifts from sexuality and pleasure to the larger questions of love and falling in love, a whole range of additional factors has to be taken into account. Love and the family bring the Oedipal complex back to the center of the picture, and love relationships (whether the object is of the same or opposite gender) have to be understood in terms of transference, ego ideals, and the splitting of the ego [24]. The coming together of the sexual and affectionate currents in a mature love relationship raises all sorts of difficulties, but failures in this coming together tend to result in what might more properly be called 'neurotic' love than 'perverse' love (e.g. Oepidal dependence or triangles are recreated, or needs for degraded or forbidden objects with accompanying patterns of psychical impotence emerge – see [11], p. 200 and reference at 200 N. 2).

[8] Or is the point (at [28], p. 101) that for heterosexuals the anus is represented by the vagina (that is, heterosexual intercourse involves displaced anal erotism)? It might for some purposes be helpful to maintain the distinction between inversion and perversion. For it then becomes easier to ask whether it is their inversion (in object) that makes some individuals perverse (in aim), or whether it is their perversion (in aim) that makes some individuals inverted (in their choice of object). Or, to put it slightly differently, the question of perversion may be relatively independent of the question of choice of object (of homosexuality or heterosexuality).

[9] Indeed, some analysts, such as Michael Balint, insist that many forms of homosexuality "are definitely not survivals of infantile forms of sexuality but later developments" ([21], p. 136). But it must be noted that many of Balint's views are insupportable, or at any rate not provided with support. In particular, of homosexuals he claims "they all know – that, without normal intercourse, there is no real contentment." (p. 142).

The deeper problem raised by lesbianism (presuming that everyone starts with a female primary love object) may be how anyone (female or male) can love a man. Is it the sameness or the maleness of the object that matters for a homosexual? Again, how does maleness matter for women? For anyone?

[10] Among the mechanisms of homosexual object-choice considered by Freud, the main one involves identification with the mother ([11], p. 145n.; [16], pp. 98–101; [24], p. 108; [25], pp. 230–231) and a secondary one involves reaction-formation against sibling rivalry ([25], pp. 231–232). Freud speaks elsewhere, in connection with a case of lesbianism, of "retiring in favour of someone else" ([23], p. 159n.).

[11] Nonetheless, at the risk of redundancy, it should perhaps be noted that there remains one difficult early passage in which Freud connects homosexuality with a transitional phase of narcissism. Writing in 1911 of the psychotic Dr. Schreber: "Recent investigations have directed our attention to a stage in the development of the libido which it passes through on the way from auto-erotism to object-love. This stage has been given the name of narcissism. What happens is this. There comes a time in the development of the individual at which he unifies his sexual instincts (which have hitherto been engaged in auto-erotic activities) in order to obtain a love-object; and he begins by taking himself, his own body, as his love-object, and only subsequently proceeds from this to the choice of some person other than himself as his object. This half-way phase between auto-erotism and object-love may perhaps be indispensable normally; but it appears that many people linger unusually long in this condition, and that many of its features are carried over by them into the later stages of their development. What is of chief importance in the subject's self thus chosen as a love-object may already be the genitals. The line of development then leads on to the choice of an external object with similar genitals – that is, to homosexual object-choice – and thence to heterosexuality. People who are manifest homosexuals in later life have, it may be presumed, never emancipated themselves from the binding condition that the object of their choice must possess genitals like their own; and in this connection the infantile sexual theories which attribute the same kind of genitals to both sexes exert much influence" ([18], pp. 60–61).

This may account for *one* type of homosexual object-choice (perhaps characteristic of

Leonardo – see Freud's study ([16], pp. 98–101), but, again, narcissism should not be confused with homosexuality. Loving oneself is not the same as loving someone else of the same gender, even if the first may in some cases lead to the second. So even if narcissism is a stage in development, homosexual object choice is not thereby reduced to such a stage.

There is a perhaps more troubling reading of the passage. Laplanche and Pontalis write: "In his first attempts to work out the idea of narcissism, Freud makes the homosexual narcissistic choice into an interim stage between narcissism and heterosexuality: the child is said to choose an object initially whose genital organs resemble its own" ([33], p. 259).

While the description in the passage may apply to some homosexuals (such as Schreber), who go on to become heterosexuals, there is no suggestion that homosexuality is a step in the standard route to heterosexuality; and Freud certainly makes no such suggestion elsewhere. Moreover, as Laplanche and Pontalis argue, "the idea of the narcissistic choice is not a straightforward one even in the case of homosexuality: the object is chosen on the model of the little child or adolescent that the subject once was, while the subject identifies with the mother who used to take care of him" ([33], p. 259).

And Freud suggests that the initial heterosexual object-love may be preserved in the process of identification: "By repressing his love for his mother he preserves it in his unconscious and from now on remains faithful to her" ([16], p. 100). And finally, as Laplanche and Pontalis conclude, "it is doubtful whether an antithesis between the narcissistic and the anaclitic object-choices, even as ideal types, is tenable. It is in 'complete object-love of the attachment type' that Freud observes 'the marked sexual over-valuation which is doubtless derived from the child's original narcissism and thus corresponds to a transference of that narcissism to the sexual object'" ([20], p. 88). Conversely, he describes the case of 'narcissistic women' in the following terms: "Strictly speaking, it is only themselves that such women love with an intensity comparable to that of the man's love for them. Nor does their need lie in the direction of loving, but of being loved; and the man who fulfils this condition is the one who finds favour with them" ([20], p. 89). It may be asked whether a case such as this, described here as *narcissistic*, does not display a subject seeking to reproduce the child's relationship to the mother who feeds it – an aim which according to Freud is a defining characteristic of the *anaclitic* object-choice" ([33], p. 259).

[12] The basic facts are recounted in [34]. A more detailed journalistic account is available in [3].

[13] This may conflict with the APA's own general characterization of a mental disorder: "a mental disorder is conceptualized as a clinically significant behavioral or psychologic syndrome or pattern that occurs in an individual and that typically is associated with either a painful symptom (distress) or impairment in one or more important areas of functioning (disability). In addition, there is an inference that there is a behavioral, psychologic, or biologic dysfunction, and that the disturbance is not only in the relationship between the individual and society. When the disturbance is limited to a conflict between an individual and society, this may represent social deviance, which may or may not be commendable, but is not by itself a mental disorder" ([5], p. 363).

C. Culver and B. Gert [30] raise difficulties of their own with the APA definitions and classifications of mental disorders, but they are less troubled than they ought to be about the category of 'ego-dystonic homosexuality'. They write: "the primary reason why certain recurring sexual behaviors are maladies is that they are ego-dystonic. The person engaging in the behavior is distressed by it. Of course, such behavior is probably also a manifestation of a volitional disability, but even if it is not, the distress, if significant, is sufficient to make it count as a malady. Note that neither in the case of distress nor of a volitional disability is the sexual condition a malady because it is sexual, but rather because of some other characteristic attached to the condition. Thus, we believe that when homosexuality qualifies as a malady it is because of the distress the person experiences, not because of the person's homosexual phantasies or desires" ([30], p. 104).

But I believe that by their own criteria for what counts as a 'malady' they should be more equivocal. They argue ([30], pp. 95–98) that grief should not be regarded as a disease because it has a 'distinct sustaining cause' (namely, an external loss – if the sufferer came to believe the loss was not real, grief and suffering would cease). And so it would seem that it is unclear whether 'ego-dystonic homosexuality' is, in their terms, a 'malady'. Doesn't the suffering (and even the putative 'volitional disability') have a 'distinct sustaining cause'? After all, if society changed its attitude, the suffering might disappear and there might be no need to overcome desires. Culver and Gert at one point write: "If a person is suffering or at increased risk of suffering evils principally because of conflict with his social environment, then his social environment would be a distinct sustaining cause of his suffering and he would not have a malady" ([30], p. 94). A theory of the source of suffering is needed if suffering is to be the sign of a malady. Even supposing a change in social attitudes would not in a given case remove suffering, when a desire is ego-dystonic, it may be because the individual has internalized mistaken standards. Is the problem then in the desire or in the standards (it is the two together that produce the distress)? Which should be changed? An individual can suffer from an unjustified (but perhaps socially encouraged) self-loathing.

[14] For example: "It is well known that a good number of homosexuals are characterized by a special development of their social instinctual impulses and by their devotion to the interests of the community. . . . the fact that homosexual object-choice not infrequently proceeds from an early overcoming of rivalry with men cannot be without a bearing on the connection between homosexuality and social feeling" ([25], p. 232).

The more usual connection that Freud makes is, of course, between social feeling and sublimated homosexuality (rather than active homosexuality): "After the stage of heterosexual object-choice has been reached, the homosexual tendencies are not, as might be supposed, done away with or brought to a stop; they are merely deflected from their sexual aim and applied to fresh uses. They now combine with portions of the ego-instincts and, as 'attached' components, help to constitute the social instincts, thus contributing an erotic factor to friendship and comradeship, to *esprit de corps* and to the love of mankind in general. How large a contribution is in fact derived from erotic sources (with the sexual aim inhibited) could scarcely be guessed from the normal social relations of mankind. But it is not irrelevant to note that it is precisely manifest homosexuals, and among them again precisely those that set themselves against an indulgence in sensual acts, who are distinguished by taking a particularly active share in the general interests of humanity – interests which have themselves sprung from a sublimation of erotic instincts" ([18], p. 61).

[15] Foot fetishism is not generally regarded as disgusting. What is disturbing or troubling about it is the idea that someone might be (sexually) interested *only* in feet. However much such focus might simplify life, it does seem to leave out other valuable possibilities.

[16] The problem here is rather like the problem with certain other behaviorist attempts to explain complex psychological phenomena. For example, Wolpe and Rachman suggest, in

relation to Freud's case of Little Hans, "that the incident to which Freud refers as merely the exciting cause of Hans' phobia was in fact the cause of the entire disorder" ([45], p. 216). The incident involved was Hans' witnessing the fall of a horse that was drawing a bus. Aside from other problems with their account (see [37], pp. 124–135), Freud had pointed out fifty years before: "Chronological considerations make it impossible for us to attach any great importance to the actual precipitating cause of the outbreak of Hans's illness, for he had shown signs of apprehensiveness long before he saw the bus-horse fall down in the street" ([14], p. 136).

Later additions to the psychoanalytic theory of fetishism (including emphasis on phases of development earlier than the phallic stage) are traced in Phyllis Greenacre, 'Fetishism' [31].

[17] I, like Nagel [35], wish to give special emphasis to the role of desires in perversion. For whether a particular activity or practice as engaged in by a particular individual should be regarded as perverse typically depends on the desires that inform his practice (though the force of this point might vary with alternative criteria for perversion and for sexuality). Description, here as elsewhere, is theory-laden. Whether a particular observable action counts as 'neurotic' depends on why it was done, on its meaning. A person who washes his hands fifteen times a day need not be obsessive-compulsive, he may be a surgeon. Similarly, a 'golden shower' performed out of sexual interest has a very different significance in respect to the question of 'perversion' than one done as an emergency measure to treat a sea urchin wound. Of course, actions can be over-determined, motives can be mixed, and motives can be hidden. In any case, the full description of what a person is doing typically depends on what he thinks (whether consciously or unconsciously) he is doing and why. Underlying thoughts and desires are essential in characterizing the nature of activities and practices.

And again, in understanding the nature of desires themselves, the role of thoughts can scarcely be overemphasized. As Stuart Hampshire concludes in the course of a discussion of the role of thought in desire: "the traditional scheme, which distinguishes the lusts from thoughtful desires, may turn out to be much too simple, and to reflect too grossly simple moral ideas. Any study of sexuality shows that thought, usually in the form of fantasy, enters into a great variety of sexual desires, which are normally also associated with physical causes. The traditional equation of physical desire, or lust, with unthinking desire is not warranted by the evidence. Nor is it true that the more reflective and fully conscious desires, which are in this sense rational, are necessarily or always the most complex. On the contrary, there can be pre-conscious and unconscious desires which are shown to have developed from very complex processes of unreflective and imaginative thought" ([32], p. 137).

[18] As Freud puts it in discussing the case of the Rat Man: "a man's attitude in sexual things has the force of a model to which the rest of his reactions tend to conform" ([15], p. 241). The thought also forms the basis for Freud's main doubt about masturbation: "injury may occur through the laying down of a *psychical pattern* according to which there is no necessity for trying to alter the external world in order to satisfy a great need" ([19], pp. 251–252; cf. [13], pp. 198–200). We should perhaps note that he continues: "Where, however, a far-reaching reaction against this pattern develops, the most valuable character-traits may be initiated."

BIBLIOGRAPHY

1. Baker, R. and Elliston, F. (eds.): 1975, *Philosophy and Sex*, Prometheus Books, Buffalo.
2. Balint, M.: 1965, 'Perversions and Genitality', *Primary Love and Psycho-analytic Technique*, Tavistock Publications, London.
3. Bayer, R.: 1981, *Homosexuality and American Psychiatry: The Politics of Diagnosis*, Basic Books, New York.
4. Boswell, J.: 1980, *Christianity, Social Tolerance, and Homosexuality*, University of Chicago Press.
5. DSM–III: 1980, *Diagnostic and Statistical Manual of Mental Disorders*, 3rd ed., American Psychiatric Association, Washington, D.C.
6. Erikson, E.: 1963, *Childhood and Society*, 2nd ed., W. W. Norton and Company, New York.
7. Freud, E. L. (ed.): 1961, *Letters of Sigmund Freud: 1873–1939*, The Hogarth Press, London.
8. Freud, S.: 1895, *Project for a Scientific Psychology*, Standard Edition I.
9. Freud, S.: 1898, *Sexuality in the Aetiology of the Neuroses*, Standard Edition III.
10. Freud, S.: 1905, *Fragment of an Analysis of a Case of Hysteria*, Standard Edition VII.
11. Freud, S.: 1905, *Three Essays on the Theory of Sexuality*, Standard Edition VII.
12. Freud, S.: 1908, *Character and Anal Erotism*, Standard Edition IX.
13. Freud, S.: 1908, *'Civilized' Sexual Morality and Modern Nervous Illness*, Standard Edition IX.
14. Freud, S.: 1909, *Analysis of a Phobia in a Five-Year-Old Boy*, Standard Edition X.
15. Freud, S.: 1909, *Notes upon a Case of Obsessional Neurosis*, Standard Edition X.
16. Freud, S.: 1910, *Leonardo da Vinci and a Memory of his Childhood*, Standard Edition XI.
17. Freud, S.: 1910, *'Wild' Psycho-Analysis*, Standard Edition XI.
18. Freud, S.: 1912, *Psycho-Analytic Notes on an Autobiographical Account of a Case of Paranoia*, Standard Edition XII.
19. Freud, S.: 1912, *Contributions to a Discussion on Masturbation*, Standard Edition XII.
20. Freud, S.: 1914, *On Narcissism*, Standard Edition XIV.
21. Freud, S.: 1915, *Instincts and their Vicissitudes*, Standard Edition XIV.
22. Freud, S.: 1917, *Introductory Lectures on Psycho-Analysis*, Standard Edition XVI.
23. Freud, S.: 1920, *The Psychogenesis of a Case of Homosexuality in a Woman*, Standard Edition XVIII.
24. Freud, S.: 1921, *Group Psychology and the Analysis of the Ego*, Standard Edition XVIII.
25. Freud, S.: 1922, *Some Neurotic Mechanisms in Jealousy, Paranoia and Homosexuality*, Standard Edition XVIII.
26. Freud, S.: 1924, *The Economic Problem of Masochism*, Standard Edition XIX.
27. Freud, S.: 1927, *Fetishism*, Standard Edition XXI.
28. Freud, S.: 1933, *New Introductory Lectures on Psycho-Analysis*, Standard Edition XXII.
29. Freud, S.: 1954, *The Origins of Psycho-Analysis: Letters to Wilhelm Fliess, Drafts and Notes, 1887–1902*, Imago, London.

30. Gert, B. and Culver, C.: 1982, *Philosophy in Medicine: Conceptual and Ethical Issues in Medicine*, Oxford University Press.
31. Greenacre, P.: 1979, 'Fetishism', in I. Rosen (ed.), *Sexual Deviation*, 2nd ed., Oxford University Press, pp. 79–108.
32. Hampshire, S.: 1975, *Freedom of the Individual*, 2nd ed., Chatto and Windus, London.
33. Laplanche, J. and Pontalis, J. -B.: 1973, *The Language of Psycho-Analysis*, The Hogarth Press, London.
34. Marmor, J.: 1980, 'Epilogue: Homosexuality and the Issue of Mental Illness', in J. Marmor (ed.), *Homosexual Behavior: A Modern Reappraisal*, Basic Books, New York, pp. 390–401.
35. Nagel, T.: 1969, 'Sexual Perversion', *The Journal of Philosophy*, **66**, pp. 5–17. Included in his *Mortal Questions*, Cambridge University Press, 1979, and in [1] and [41].
36. Nagel, T.: 1976, 'Moral Luck', *Proceedings of the Aristotelian Society*, vol. Supp. L, pp. 137–151. Included in his *Mortal Questions*, Cambridge University Press, 1979.
37. Neu, J.: 1977, *Emotion, Thought, and Therapy*, Routledge & Kegan Paul, London.
38. Neu, J.: 1981, 'Getting Behind the Demons', *Humanities in Society* **IV**, 171–196.
39. Ruddick, S.: 'Better Sex', in [1], pp. 83–104.
40. Slote, M.: 'Inapplicable Concepts and Sexual Perversion', in [1], pp. 261–267.
41. Soble, A. (ed.): 1980, *The Philosophy of Sex: Contemporary Readings*, Littlefield, Adams and Co.
42. Taylor, C.: 1964, *The Explanation of Behaviour*, Routledge & Kegan Paul, London.
43. Williams, B.: 1976, 'Moral Luck', *Proceedings of the Aristotelian Society*, vol. Supp. L, pp. 115–135. Included in his *Moral Luck*, Cambridge University Press, 1981.
44. Wilson, E. O.: 1978, *On Human Nature*, Harvard University Press.
45. Wolpe, J. and Rachman, S.: 1963, 'Psychoanalytic Evidence: A Critique Based on Freud's Case of Little Hans', in S. Rachman (ed.), *Critical Essays on Psychoanalysis*, Pergamon, Oxford, pp. 198–220.

SANDRA HARDING

THE POLITICS OF THE NATURAL: THE CASE OF SEX DIFFERENCES

THE DISAPPEARANCE OF NATURAL SEX DIFFERENCES

It is difficult these days to talk intelligibly about what is really natural or unnatural, normal or abnormal, about sex differences, human sexuality or the human body. The distinction between natural/unnatural or normal/abnormal presumes a standard for what is given us by biology and what by culture. But contemporary feminists argue convincingly that women, like men, appear in everyday life as a socially constructed sexual *class*, not primarily as a biologically distinct group. Of course males inseminate and females incubate and lactate. There are male and female developmental processes that account for this reproductive difference and are defined in terms of five biological criteria: genes or chromosomes, hormones, gonads, internal reproductive organs, and external genitalia ([40], p. 11). However, behavioral differences between the sexes overwhelmingly appear to be the consequence of relevant social differences. As Simone de Beauvoir wrote:

> One is not born, but rather becomes a woman. No biological, psychological, or economic fate determines the figure that the human female presents in society: it is civilization as a whole that produces this creature, intermediate between male and eunuch ([12], p. xiv).

Though traditionally men's behaviors and activities have never been thought to be as biologically constrained as women's, the same argument supports the insight that one is also not born a man but rather becomes one.

In the second place, these feminist arguments are supported by the findings of sex researchers and psychoanalytic theorists. Human sexuality is plastic, not rigidly controlled by genetic or hormonal patterning [6, 40, 41, 60]. Human infants are born bisexual or 'polymorphously perverse', in Freud's phrase. Research on the sexual identity of hermaphrodites shows complete disjunction between the physiological sex of the hermaphrodite infant (defined by the five criteria mentioned) and

E. E. Shelp (ed.), Sexuality and Medicine, Vol. I, 185–203.

the eventual erotic/sexual identity adopted by the child. Parental expectation, not physiological sex, predicts the eventual sex identity of the infant [40]. What is true for these abnormal cases (between 2 and 3% of humans are estimated to be hermaphrodites) also appears to be true for the rest of us: social expectation produces sexual/erotic identity [11, 13].

Finally, post-Kuhnian social studies of science cast further doubt on the idea of a nature that is 'out there' for individual scientific geniuses to discover in the course of their pursuit of truth. It is increasingly difficult to continue to hold to the Enlightenment belief that science is a mirror of a pre-social nature [49]. To the contrary, studies by Jerome Ravetz, Alfred Sohn-Rethel, David Bloor, Barry Barnes, Bruno Latour and Steve Woolgar and others show science *in* society, and scientists' hopes for society projected into even the most abstract elements of science [1, 5, 10, 32, 47, 50, 57]. These studies make it impossible to maintain the image of the scientist as a craft laborer creating his individualized 'work' of impartial knowledge when the sociological reality of science production in advanced industrial societies is one of a process requiring the participation of literally millions of interchangeable wage laborers under the direction of a scientific bureaucracy highly integrated into state and bourgeois projects [1, 50]. Feminist scholars have contributed to this perspective by pointing to the distinctively masculine bias in the selection and definition of what is scientifically problematic, in the dominant concepts and theories in the particular sciences, in the design of research projects, in the interpretation of the results of research, and in the social organization of science [19, 23, 24, 25, 28, 39, 53, 55, 59].

These considerations should dissolve the question of identifying the 'natural' in human sexuality as well as in every other aspect of humans and our surroundings. Our dominant concepts of nature and of inquiry are in fact created by identifiable historical projects. There are no 'natural bodies' detectable in the dominant scientific conceptual schemes. Nowhere can we find nature itself, untouched by social life, in the concepts, theories, and social practices of either a hierarchically organized social order or of its science. Indeed, these critics argue that the very concept of 'natural phenomena' is empirically meaningful only as one part of an identification of other phenomena as the cultural or unnatural: science itself invented the modern form of this dichotomy. Furthermore, scientific culture consistently has associated the social labors of women, people of color, and 'wage-slaves' with the natural against which European men's intellectual and administrative activities are defined as the distinctively cultural, the distinctively human. The referents for 'nature' and 'sexed bodies' even shift from era to era within the last four centuries of Western social life, and these terms often have incompatible referents even within the same epoch [9, 27]. These criticisms lead to a cultural constructionist understanding of both sexuality and of science (not necessarily a cultural determinist one).

In light of these recent critiques, their widespread dissemination, and the consequent scientific and popular strengthening of cultural constructionist approaches, it takes an intellectual heroism of sorts to assume that one can continue to discuss what is natural or normal about human sex differences. But such heroes and heroines of the natural order we have among us. In this essay I will review central conceptual obstacles to continued attempts to naturalize sex differences, point to the contradictory public agendas defended by proponents of natural sex differences, and conclude with some implications of this issue for medical approaches to sexuality.

SOME OBSTACLES TO NATURALIZING SEX DIFFERENCES

Both biological determinist and cultural constructionist perspectives on sex differences appeal to scientific understandings in support of their claims. In different ways, both use rhetoric stressing sameness and difference. The biological determinists focus on the fact that it is the sex differing developmental cycles mentioned earlier that make it possible for male insemination and female incubation and lactation. Science reveals, they claim, that human females and males are fundamentally, 'essentially', different from each other and like their counterparts in other species. By stressing the fundamentalness of sex difference within the species and sexual sameness across species, they justify attributing to reproductive difference a vast array of behaviors and activities segregated by sex in our culture, but not easily accounted for as the direct effect of reproductive differences.

The cultural constructionists appeal to scientific understandings in support of their claim that there is far greater variation in sex differing behaviors in our species than in others. It is the differences between our species and others that must be theorized in order to account for this cultural variation. They point out that what is biologically significant about males and females in our species is that they share this innate and inevitable plasticity. Human nature is innately, inevitably, and 'essentially' immensely more plastic, more adaptable and inventive than 'orangutan nature' or 'termite nature'.

Paradoxically, the scientific claims to which cultural constructionists appeal can also be understood as biological determinist. Humans are biologically determined to invent ourselves and our forms of interaction with our social/natural surroundings, and no account of 'false consciousness' denying our differences from other species makes this less true. Insisting on, arguing for, or assuming severe biological constraints on human behaviors and actions are distinctively human acts – they are choices arrived at through custom or deliberation, but ones not available to members of other species. Philosophers will be reminded of Sartre's insight that humans are 'condemned to be free'. Though it is important to remember that both sides to this dispute appeal to biology in support of their claims, I shall follow popular usage and refer only to those who emphasize our similarities to other species as 'biological determinists'.

There are conceptual obstacles that appear to limit our ability unambiguously to identify biological components in sex differing behaviors beyond the 'minimum' directly associated with the fact that males inseminate and females incubate and lactate. Quite apart from the lack of empirical evidence, there is a lack of conceptual clarity, making biological determinist claims about non-reproductive behavior questionable. In outlining these problems, I do not presume to enter the intricacies of empirical scientific argument – intricacies appropriately addressed by scientists themselves. The objections I review are philosophical ones; they are objections to the kinds of concepts and forms of explanation that must be relied on in attributing biological causes to sex differing behaviors beyond what is necessary to account directly for reproductive differences.

The first problem is that inspection of neither the biological correlates nor the prevalance of sex-differing behaviors in themselves permits the isolation of biological from cultural components. If certain 'life-styles' can be correlated with the incidence of ulcers, strokes, heart attacks, and nutrition related developmental and disease phenomena, obviously we are not forced to assume that observable physiological differences between humans can be understood in isolation from the cultural environments in which these phenomena occur. Thus when hormonal researchers argue that the higher level of testosterone in males causes male 'aggression' and male 'dominance', and consequently accounts for all of the important contributions to the creation of civilization, they assume what needs explanation. First of all, the terms 'aggression' and 'dominance' in these studies are permitted to range selectively over such diverse and not obviously related phenomena as willingness to fight, access to females, 'grooming hierarchies', physical size, verbal displays, 'acknowledgements of precedence', competitiveness, etc. [8, 33]. In the second place, what is often discussed as 'dominance' behavior among primates (and humans?) may in reality be 'subordinance' behavior, for the 'approach behaviors' which researchers often take as signs of high dominance status appear to reflect the stress of subordinance ([54], cited by [33], p. 44). Thus, finally, it also is not clear whether high testosterone levels cause any of these aggressive behaviors, or whether aggressive behaviors can cause high testosterone levels ([8], [37], p. 107).

Similar obstacles arise to asserting that the fact of the prevalence of sex-differing behaviors reveals these behaviors to be adaptive and thus to have uniquely biological origins. As sociobiologists themselves have pointed out, " . . . just because something is adaptive does not mean that it must be biological even if it accords with sociobiologic prediction . . . social and cultural factors may mimic the action of natural selection regarding sex differences in behavior" ([4], p. 283, quoted by [38], p. 95). Thus even if we could identify a sex differing behavior that has persisted through the whole panorama of cultural history, we would not thereby have evidence that the behavior is biologically destined. (Nor does identification of cultural influences show that a trait or behavior has no biological origins.) Thus the attempt to isolate biological from cultural 'causes' of sex differing behaviors is supported neither by the identification of biological mechanisms correlated with such differences nor of patterns of difference. What we would need is a *theory* to guide us in identifying which of the many observable sex differing behaviors not directly necessary for reproduction might nevertheless have uniquely biological causes.

However, there is at present no viable theory of such differences, and consequently no way to sort relevant from irrelevant observed differences. In the absence of such a theory, observation of differences does not constitute evidence for or against any hypothesis. The lack of viable theories of the biological bases of sex differing behaviors not directly necessary to achieve reproduction makes this field of research like the fact-collecting natural histories of the 17th century which "juxtapose facts that will later prove revealing . . . with others . . . that will for some time remain too complex to be integrated with theory at all" ([29], p. 16). Without a viable theory, all differences become equally interesting. With a theory, we can hypothesize which kinds of differences are linked in 'natural patterns' and which are not. The conclusion of a comprehensive review of research on sex differences in cognition ends by pointing out that "It is clearly very easy to include sex as a bonus factor in experiments which have little scientific merit. We cannot pretend that we are testing a theory of sex differences, since at present none can exist" ([18], p. 280, quoted by [38], p. 28). Consequently,

> . . . the majority of the studies reviewed here and elsewhere are both ill-thought and ill-performed. Whilst in other circumstances this may be regarded as the occupational hazard of the scientific enterprise, here such complacency is compounded by the social loadings placed upon these kinds of results [Ibid.].

Why is it that no such theory at present 'can exist'? It appears that the search for such a theory is conceptually confused. Genetic inheritances constitute arrays of possibilities, and which of these possibilities will be expressed depends on the environment within which the genes are located. For humans, our genetic 'environments' are culturally selected as well as culturally constructed. "Behavior results from the joint operation of genes and the environment, and these factors interact in complex and nonlinear ways that are different and unpredictable for different traits" ([38], p. 95). Thus it is meaningless to try to partition genetic and environmental components in behaviors and discuss them separately.

> In the absence of knowledge about these interactions for each specific trait one wishes to consider – and at present we do not have it about a single observed human behavior – the only meaningful question that can be asked is how much of the observed *variation* in behavior among individuals is caused by genetic factors and how much by environment. . . . This more limited question tells one *nothing* about how to partition genetic and environmental effects for the behavior itself; nor does it tell us anything about the proportion of genetic and environmental contributions to the variance of any other trait ([38], pp. 95–96).

Furthermore, even if we did manage to partition the variation for any particular trait, this feat would not enable us to predict that the same partition would appear in a different environment: the relative contributions of genetic and environmental factors are likely to change when the environment changes. "Thus, in comparing two groups that differ genetically, it is impossible to distinguish the genetic and environmental origins of *any* behavioral differences between them as long as their environments differ in *any* way" ([38], p. 95).

Finally, these conceptual obstacles are supported by a different kind of constraint: the fact that our species differs from others in its relative plasticity. Claims that do not account for this plasticity fail to account for uncontroversial empirical facts. One biologist explains this constraint this way:

> Even if some sex-associated behaviors *were* found to be universal among all nonhuman primates or indeed among all mammalian species, generalizations to human behavior and social relationships would have to ignore five million years of exuberant evolutionary development of the human brain, which has resulted in a cerebral cortex quantitatively and qualitatively different from that of other primates. It is a cortex that provides for conceptualization, abstraction, symbolization, verbal communication, planning, learning, memory and association of experiences and ideas, a cortex that permits an infinitely rich behavioral plasticity and frees us, if we choose, from stereotyped behavior patterns. . . . Not only is there no universal behavioral trait or repertoire among our closest relatives, the nonhuman primates, to study as a "primitive" prototype or precursor model for human "nature", there is no *human nature*, no universal human behavioral trait or repertoire that can be defined, *except* for our tremendous capacity for learning and for behavioral flexibility ([8], pp. 58–59).

In summary, there are fundamental conceptual obstacles limiting the fruitfulness of attempts to explain the vast array of non-reproductive sex differing behaviors as biologically determined.

DIFFERENCE AND SAMENESS AS POLITICAL RESOURCES

In light of these obstacles to establishing biological bases for sex differing behaviors beyond 'the reproductive minimum', the selective political uses of these sameness/difference contrasts reveal an illuminating pattern. Appeal to perceptions of natural sex difference constitute scientific and political resources that are used in different ways by three groups with radically different political agendas in the history of the ongoing battle between the sexes. Furthermore, two of the groups are self-proclaimed feminists and reasonably regarded as such. Thus biological determinism does not divide feminists from their opponents in some simple way. All three appeals to natural sex differences are responses to the focus on sexual sameness and species difference that has characterized mainstream feminism and some sectors of radical feminism.

In periods of accelerated social change, women and their male supporters find the resources and opportunities to form movements through which to agitate for reforms increasing women's status relative to men's. The central public agenda struggles in these campaigns are organized around assertions of sexual sameness. Mainstream feminists point out that women are like men in that they possess the capacities and abilities to vote and to hold positions of political leadership, to hold 'men's jobs' and receive equal wages for their labor, to gain access to the education and credentialling heretofore reserved for men, and otherwise to participate as equals in political, economic, and moral decisions about their own lives and the lives of others. These feminists argue that there are no significant differences in the 'natures' of women and men sufficient to account for women's continuing exclusion from full participation in the design and maintenance of the institutions of government, the economy, education, or moral and religious values – four of the five social institutions, as sociologists point out, requisite to every culture. (The fifth institution, notably missing from this list of arenas in which women and men are claimed to be 'the same', is of course the family.) These were the arguments of the 19th century women's movement, and they have reappeared again in the 1960s and 1970s.

When gains are made through feminist agitation and other forces of social change, proponents of sexual sameness find themselves opposed by the proponents of sexual difference. These difference defenders can be classified into three groups. Most obvious are *Reactionary Patriarchs*, the men who have been quite comfortable controlling these heretofore male preserves, and who are the main target of feminist agitation. But there are also two groups of feminists who object to the sexual sameness arguments. There are the *Revisionist Feminists*: women and men who want equality of social recognition for traditionally feminine values and activities, who fear that these are being devalued now by women themselves through the mainstream feminist agitation to make women just like men in the public realm, but who are less willing to challenge the sexual order in the public sphere. These Revisionist Feminists support a separate but equal political agenda. Finally, there are the *Feminist Radicals*: those who think women's traditional talents and abilities should neither be devalued nor restricted to women's traditional spheres. They think that everyone should try to become more like women. To some extent these labels caricature the positions of the individuals who defend these views; but they enable us to distinguish the political agendas of the heroes, heroines, and amazons who often appeal to the naturalness of sex difference in reaction to the claims of sexual sameness advanced by mainstream feminism.

Let us further consider the pattern of appeal to the naturalness of sex differences by current participants in the battle between the sexes. In the first place, there is the reactionary right. These defenders of male dominance argue that the fact of reproductive difference between the sexes necessitates as natural, normal, and morally desirable 'the biological family', heterosexuality, masculine and feminine gender, and the separation of the public from private life (by which they mean the continued exclusion of women from positions of authority in public life and the right of men to bear few of the costs or responsibilities for family life). They argue this with respect to a wide array of phenomena in which the psychiatric and medical professions are confronted with the vagaries of human sexuality – abortion, sterilization, contraception, marital and non-marital rape, spouse abuse, child abuse, child care, social services for homosexuals, marital unhappiness, 'deviant' expressions of sexual desire, etc. By masculine and feminine gender, the reactionary right means male dominance and female submissiveness. Without this extensive sexual differentiation, they fear that 'civilization' – by which they mean modern, Western, racist, classist and sexist civilization – will fall. Examples of these arguments can be found in the writings of Konrad Lorenz, Desmond Morris, Robert Ardrey, Lionel Tiger, E. O. Wilson, George Gilder, and Michael Levin, to mention some of the more widely-known Reactionary Patriarchs [2, 22, 35, 36, 42, 61, 64].

The second appeal to natural sex differences is made by those their opponents label 'revisionist feminists' [14] or 'conservative feminists' [58]. Like the Reactionary Patriarchs, these writers also believe that the fact of reproductive difference between the sexes requires as natural, normal, and morally desirable 'the biological family', heterosexuality, masculine and feminine gender, and the separation of the public from the private in many of the areas Reactionary Patriarchs focus on. But by masculine and feminine gender, they mean sexual 'equality' – the maintenance of separate but equal genders with separate but equal domains for their expression. The Revisionist Feminists are specifically opposed to racism, classism, and sexism. These tenets of Revisionist Feminism have recently been defended by Alice Rossi, Jean Bethke Elshtain, Betty Friedan, Peter Berger and Brigitte Berger, Christopher Lasch, and Ivan Illich [7, 15–17, 21, 26, 31, 52].

Both the Reactionary Patriarchal and the Revisionist Feminist defenders of 'the natural' perceive their own efforts as heroic last-ditch stands against escalating forces of irrationality and social disorder, which they define as emanating mainly from mainstream feminism. This conception of the nature of their project and this identification of their opponents is bizarre for several reasons. For one thing, the perspective that understands sexuality and 'natural bodies' as political constructs is far broader than mainstream feminism. As I have indicated, this is a major implication of the post-Kuhnian academic social studies of science and of psychoanalytic and sex theory researchers. It is also a major thesis for a diverse group of contemporary thinkers sharing no feminist motivations at all and including Michael Foucault, William Leiss, and those influenced by 'the Freudian Left' [20, 34].

Furthermore, the social disorder attributed by both Reactionary Patriarchs and Revisionist Feminists to the mainstream feminism, which emerged only in the late 1960s and early 1970s, has obvious origins that not only pre-date the second wave of American feminism but in fact made feminism possible. The plausibility of this reversed causal analysis is irresistible, once one thinks about the invention of cheap contraceptives for the purpose of controlling Third World populations and internal minorities, changes since World War II in higher education, the shift in growth sectors of the economy from industrialized 'men's work' to the service sectors traditionally employing women, of social welfare policies, the Civil Rights movement, and changes in 'the family' caused by all of these factors. Since these kinds of changes pre-date the emergence of the second wave of feminism, they cannot be caused by feminism. It is these changes that have made possible increased 'sexual sameness' and feminism, not the reverse.

The third appeal to biological determinism in defense of sexual difference is an often overlooked and somewhat fragile stream of radical feminist thinking. The argument here is that women's different reproductive system better prepares women for socially acquiring talents and skills that should be used in designing and maintaining a more emancipatory public life for all. Everyone should try to become more like women. I say this Radical Feminist argument is 'fragile' because it is difficult to find single authors who explicitly move from the biological premises to the political conclusions. Nevertheless, I feel justified in pointing to the existence of this at least 'ideal type' Radical Feminist defense of sexual difference, since it can easily be reconstructed from anthropological and historical speculations about how gender got created and the sexual order ever got started in the first place, as well as from psychological hypotheses about the relation of sexual embodiment to the social construction of sexual identity for infants. Here appeals to natural sex differences function as a resource for arguments for the most radical of social reforms.

This argument begins in the recognition that in many cultures sexual symbolism focuses on ways in which the boundaries of women's bodies are less clear than for men's. In menstruation, women bleed but do not die. In intercourse, women's bodies are entered for mutual pleasure. In pregnancy, another human lives inside a woman's body and shares her biological systems. In nursing, women's bodies become food for other humans. Adrienne Rich writes of the difficulty for the pregnant woman to discern where her body ends and the fetus's begins [48]. Two streams of post-Freudian psychoanalytic theorists, object relations theorists and Lacanians, describe the oneness between mother and infant created by women's greater interaction with infants during nursing – a oneness the 'law of the father' must interrupt to entice the child into both gender and individual identity [11, 13, 30]. The political theorist Mary O'Brien describes female and male 'reproductive consciousness' created through the different reproductive participation of males and females [44]. The argument made by all of these authors is that biological differences between the sexes predispose women to insertion into social roles of mediating the boundaries between nature and culture [45]. Since our culture suffers from an institutionalized insensitivity to 'nature's' own ways, exhibited in the constant and dangerous attempt to dominate nature and other people, it is women's unique sensitivities that should be infused into public life in an appropriately modified form (i.e., not as the forms of the feminine created through male dominance, but as the truly human), and that are already appreciated in the ecology movement, anti-militarism, and agitation for social welfare [3]. Some of these Radical Feminists think that since bottle-feeding makes unnecessary a greater involvement by women than men in infant care, gender and the sexual order can be eliminated if men are equally involved in infant care and women in 'ruling' in the public sphere (*and* if no other social forces, such as a capitalist economic order, are investing in male dominance and female submission, [11, 13]). Some think that co-parenting is the single most important step to take in eliminating not only sexism but also classism [3]. Some think that co-parenting is irrelevant to the formation of gender, sexual identity and the sexual order; that 'the law of the father' creating all three is directly a consequence of being born to women – creatures lacking a penis/phallus [30]. Yet others think that it is the revolution in birth control technology that will make it possible for women to end the association of women with nature and men with culture [44].

Common to these various arguments is the claim that, at least until now, reproductive differences between the sexes have predisposed women to the development of abilities to mediate between culture and nature, that contemporary Western culture is mutilating to its members and dangerous to life on this planet because it is not guided by these mediating sensitivities, and that men should be encouraged to develop these mediating sensitivities to which women's biology predisposes them. The sources for this argument differ in their assessment of which (if any) of the changes in the social organization of reproduction made possible by bottle-feeding and birth control are the ones that would bring about the elimination of gender and a male dominant sexual order; but all agree that so far reproductive differences between the sexes have played a causal role in the social production of tendencies we can now identify as leading directly toward widespread social disaster.

Appeals to natural sex differences for the same three public agendas can be found in the 18th and 19th centuries. Reactionaries argued that women's reproductive systems would be endangered and civilization would fall if women gained access to public education, economic independence, and suffrage [51, 53]. Defenders of 'republican motherhood', the 19th century forerunners of today's Revisionist Feminists, argued that women's natural mothering skills and the higher moral sensibility thereby developed were a valuable social resource justifying education for women – but only so that they might perform more enlightened mothering of the next generation of citizens [51]. Finally, Rousseau and other French social reformers criticized the evils of man-made culture.Their arguments implied that women's greater access to 'the natural' – women's condition as mediator between nature and culture – should lead to women, not men, designing and ruling an enlightened public life [9]. Of course, when this logical consequence of their exaltation of a 'nature' they consistently associated with women became clear to them, they revised their demands for social reform to block this unfortunate implication of their critique of unnatural culture [9].

Groups with wildly different agendas for public policy are all now appealing to biological sex differences in support of their arguments, and the pattern visible today replicates patterns of appeal in the past.

MEDICINE AND SEXUALITY

There are a number of points at which this discussion of the political uses of appeals to natural sex difference is relevant to medical approaches to sexuality.

In the first place, science and medicine have been deeply involved in perpetuating the delusion that we can identify distinctively biological constants beneath the flamboyant variation in observed human traits and behaviors. The very concept of 'natural bodies' is the product of science's and medicine's attempts both to model all inquiry on physics and to legitimate their status to a public entranced with the technologies science makes possible and anxious about the constant changes in social relations characteristic of distinctively modern cultures. But there is nothing inherent in science or medicine that requires their continued contribution to the delusion that we can isolate 'the natural' in human biology from the cultural and environmental with respect to human sexuality or any other aspect of the phenomena around us. As noted earlier, the appeals to sexual sameness also claim science as their legitimation.

However, it must be noted that science and medicine's complicity in the isolation of 'the natural' in human sexuality proceeds in part from a certain aspect of the 'logic' of scientific inquiry coupled with the separation of the physical from the social sciences. Scientific method leads to a bias toward focus on differences. It is against an assumed background of sameness that experimental method can produce crucial differences, which provide clues to the lawful 'samenesses' creating diverse observable phenomena. When the subject matter of inquiry is biological organisms and mechanisms, rather than humans in the richness of their cultural environments, sexual differences fairly leap to the eye. The training of physicians and biologists as physical scientists leads them to devalue cultural context and to fixate on physiological differences as more 'real', more fundamental, than cultural and environmental differences. It also leads them to devalue the samenesses in human males and females. Thus the very division between the physical and the social sciences itself obscures regularities and underlying causal determinants in humans and the world around us. Why should we any longer respect this mystifying division?

In the second place, the flourishing of scientific/medical research projects focused on describing the regularities and explaining the biological causes of perceived sex differing behaviors is responsive not only – and probably not primarily – to the 'internal logic' of scientific inquiry, as the kind of positivist philosophy of science fed to fledgling scientists would lead them to believe. Undervalued in the standard philosophies of science are the roles of political perceptions and definitions of what is problematic and thus in need of scientific explanations [43, 46]. If men's heterosexuality as it is stereotypically expressed in Western post-industrial cultures is taken as 'natural' and as the norm for human sexuality, then women's sexuality, homosexuality, and the sexual desires and practices of other cultures will be seen as in need of explanation. Feminists, among others, have asked different questions. In particular, they have suggested that normal, Western, male sexuality and compulsory heterosexuality should be seen as problems requiring explanation. How does initially 'polymorphously perverse' infantile sexuality get channelled into rigidly heterosexual masculinity and femininity? Why is men's sexuality so driven, so compulsive, and so pervasive a mode of expressing social relations that appear to have nothing to do with strict reproductive needs? Why are metaphors of domineering and driven masculine sexuality standardly used in the rhetoric of militarism, race and class relations, interactions with nature? These new problems emerge not from any 'internal logic' of the 'growth of scientific knowledge', but from political agitation by those who fear the destructive impulses unleashed by the cultural exaltation of a dominant masculinism.

Correlatively, Reactionary Patriarchal and Revisionist Feminist appeals to the biological constraints on sex differences are receiving wide public attention not because of the greater excellence of the scientific foundations of these claims or because of the greater relevance of sexual difference to the design of the social order. Instead, this happens because these visions are hopeful to those at the top of socially stratified societies such as ours, and to those fearful of social changes. The Reactionaries posit a natural aristocracy that justifies policy making by and on behalf of those who already have a disproportionately large voice in public policy. The Revisionists echo this politic, if in a more muted form. In a whole array of domains where medicine and health services are presented with the vagaries of human sexual behavior, there has been increasing restriction on what can and cannot be done with public funds. Public funding is increasingly being withdrawn from abortions, contraceptive counseling, health services for homosexuals, policing of occupational health standards, rape crisis centers, and centers for battered women, as well as from an array of social institutions and policies servicing primarily indigent women and their children. This funding is now being channelled into increased military spending, an array of incentives to 'private enterprise', genetic engineering, and fetal research. Even if men were biologically destined to rape and batter, impregnate outside of marriage, even if women's 'destiny' as child bearers sometimes leaves them indigent, there is no ethical imperative to be found in the 'logic' of biological determinism that would justify deteriorating even further the condition of the 'worst off' in our society. But that is the political use to which biological determinist claims are currently being deployed. The public order is not interested in the political agenda that the Radical Feminists claim by a similar appeal to 'natural' sex differences. If appeal to natural sex differences can be used to support such contradictory public agendas, it should be clear that it is not the facts of biological difference but what we make of them that is relevant to the design of our collective life.

Third, all parties to this dispute – mainstream feminists as well as the three varieties of defenders of the natural order – conduct their discourses within a conceptual scheme emphasizing differences against a background of samenesses. Different uses of this conceptual contrast distort our understandings of human sexuality in different ways. The three appeals to biological determinism focus on difference between the sexes against the assumed background of sameness between ours and other species. This leads to distorting illusions that the social order is, can be, or should be identical to an imagined natural order and makes incomprehensible aspects of the social order, such as racism and classism, which are not 'derivable' from sex differences. The mainstream feminist appeal to sexual sameness, insofar as it occurs within the ongoing history of taking male activities as the paradigm or norm of species activities, leads to the illusion that what is different about our species is its cultural masculinity, what is the same about the sexes is that they are both capable of masculine achievement. It undervalues the distinctively cultural character and contributions of 'women's work'. This focus, too, often makes invisible the race and class differences between the men whose activities women would join as equals in the public arena. It is presumably not exploited black male workers to whose activities women want equal access. It is important that women take their rightful and socially needed place in the public sphere, but it is also crucial that racism and classism be recognized and dismantled.

Finally, the identification and definition of the scientifically problematic appears to depend as much on the presence of new voices in scientific debates as it does on the internal logic of scientific inquiry. Thus enlightened scientific self-interest would prioritize the education and research support of the kinds of social persons heretofore largely absent from scientific and medical discourse. Women, people of color, and homosexuals are the most obvious groups whose perspectives on what is 'problematic' in human sexuality and all other aspects of our collective life together promise quantum leaps in our understandings, and hence in the possibilities for designing the first truly human social relations.

University of Delaware,
Newark, Delaware,
U.S.A.

ACKNOWLEDGEMENT

I wish to thank Ann Ferguson for critical comments on earlier drafts of this paper.

BIBLIOGRAPHY

1. Arditti, R., Brennan, P., and Cavrak, S. (eds.): 1979, *Science and Liberation*, South End Press, Boston, Mass.
2. Ardrey, R.: 1966, *The Territorial Imperative*, Atheneum, N.Y.
3. Balbus, I.: 1982, *Marxism and Domination*, Princeton University Press, Princeton, N.J.
4. Barash, D.: 1977, *Sociobiology and Behavior*, Elsevier, New York.
5. Barnes, B.: 1977, *Interests and the Growth of Knowledge*, Routledge and Kegan Paul, London.
6. Beach, F. A.: 1947, 'Evolutionary Changes in the Psychological Control of Mating Behavior in Mammals', *The Psychological Review* **54**, 297–313.
7. Berger, P. and Berger, B.: 1983, *The War Over the Family: Capturing the Middle Ground*, Anchor, N.Y.
8. Bleier, R.: 1979, 'Social and Political Bias in Science: An Examination of Animal Studies and Their Generalizations to Human Behavior and Evolution', in E. Tobach and B. Rosoff (eds.), *Genes and Gender II: Pitfalls in Research on Sex and Gender*, Gordian Press, New York, pp. 49–70.
9. Bloch, M. and Bloch, J.: 1980, 'Women and the Dialectics of Nature in Eighteenth Century French Thought', in C. MacCormack and M. Strathern (eds.), *Nature, Culture and Gender*, Cambridge University Press, New York, pp. 25–41.

10. Bloor, D.: 1977, *Knowledge and Social Imagery*, Routledge and Kegan Paul, London, England.
11. Chodorow, N.: 1978, *The Reproduction of Mothering*, University of California Press, Berkeley, California.
12. De Beauvoir, S.: 1952, *The Second Sex*, Bantam, New York.
13. Dinnerstein, D.: 1976, 'The Mermaid and the Minotaur', *Sexual Arrangements and Human Malaise*, Harper and Row, New York.
14. Eisenstein, Z.: 1983, 'Feminist Revisionism: Debating Sexual Politics', unpublished manuscript.
15. Elshtain, J.: 1979, 'Feminists Against the Family', *The Nation*, Nov. 17, 481–482.
16. Elshtain, J.: 1981, *Public Man, Private Woman: Women in Social and Political Thought*, Princeton University Press, Princeton, N.J.
17. Elshtain, J.: 1982, 'Antigone's Daughters', *Democracy* **2**, 46–59.
18. Fairweather, H.: 1976, 'Sex Differences in Cognition', *Cognition* **4**, 231–280.
19. Fee, E.: 1983, 'Women's Nature and Scientific Objectivity', in M. Lowe and R. Hubbard (eds.), *Woman's Nature: Rationalizations of Inequality*, Pergamon Press, New York, pp. 9–28.
20. Foucault, M.: 1980, *The History of Sexuality*, Vol. I, Vintage Books, N.Y.
21. Friedan, B.: 1981, *The Second Stage*, Summit Books, New York.
22. Gilder, G.: 1973, *Sexual Suicide*, Quadrangle/The New York Times Book Co., New York.
23. Haraway, D.: 1978, 'Animal Sociology and a Natural Economy of the Body Politic: Parts I and II', *Signs: Journal of Women in Culture and Society* **4**(1), 21–60.
24. Harding, S. and Hintikka, M. B., (eds.): 1983, *Discovering Reality: Feminist Perspectives on Epistemology, Metaphysics, Methodology and Philosophy of Science*, D. Reidel, Dordrecht, Holland.
25. Hubbard, R., Henifin, M. S., and Fried, B. (eds.): 1982, *Biological Woman: The Convenient Myth*, Schenkman Publishing Co., Cambridge, Mass.
26. Illich, I.: 1983, *Gender*, Pantheon, N.Y.
27. Jordanova, L. J.: 1980, 'Natural Facts: A Historical Perspective on Science and Sexuality', in C. MacCormack and M. Strathern (eds.), *Nature, Culture and Gender*, Cambridge University Press, New York, pp. 42–69.
28. Keller, E. F.: 1982, 'Feminism and Science', *Signs: Journal of Women in Culture and Society* **7**(3), 589–602.
29. Kuhn, T. S.: 1970, *The Structure of Scientific Revolutions*, University of Chicago Press, Chicago, Ill.
30. Lacan, J.: 1982, *Feminine Sexuality*, J. Mitchell and J. Rose (eds.), W. W. Norton, New York.
31. Lasch, C.: 1977, *Haven in a Heartless World: The Family Besieged*, Basic Books, N.Y.
32. Latour, B. and Woolgar, S.: 1979, *Laboratory Life: The Social Construction of Scientific Facts*, Sage Publishing Co., Beverly Hills, Calif.
33. Leibowitz, L.: 1979, '"Universals" and Male Dominance Among Primates: A Critical Examination', in E. Tobach and B. Rosoff (eds.), *Genes and Gender II: Pitfalls in Research on Sex and Gender*, Gordian Press, New York, pp. 35–48.
34. Leiss, W.: 1972, *The Domination of Nature*, Beacon Press, Boston, Mass.
35. Levin, M.: 1970, 'The Feminist Mystique', *Commentary* **70**(6), 25–30.

36. Lorenz, K.: 1966, *On Aggression*, Harcourt, Brace & World, New York.
37. Lowe, M.: 1982, 'Social Bodies: The Interaction of Culture and Women's Biology', in R. Hubbard, M. S. Henifin, and B. Fried (eds.), *Biological Woman: The Convenient Myth*, Schenkman, Cambridge, Mass., pp. 91–116.
38. Lowe, M. and Hubbard, R.: 1979, 'Introduction', 'Sociobiology and Biosociology: Can Science Prove the Biological Basis of Sex Differences in Behavior?', and 'Conclusions', in E. Tobach and B. Rosoff (eds.), *Genes and Gender II: Pitfalls in Research on Sex and Gender*, Gordian Press, New York, pp. 9–34, 91–112, 143–152.
39. Merchant, C.: 1980, *The Death of Nature: Women, Ecology and the Scientific Revolution*, Harper and Row, New York.
40. Money, J.: 1965, 'Psychosexual Differentiation', in John Money (ed.), *Sex Research: New Developments*, Holt, Rinehart and Winston, New York, pp. 3–23.
41. Money, J. and Tucker, P.: 1975, *Sexual Signatures: On Being a Man or a Woman*, Little, Brown and Company, Boston.
42. Morris, D.: 1968, *The Naked Ape*, McGraw-Hill, N.Y.
43. Nagel, E.: 1961, *The Structure of Science*, Harcourt Brace and Row, New York.
44. O'Brien, M.: 1981, *The Politics of Reproduction*, Routledge and Kegan Paul, New York.
45. Ortner, S.: 1974, 'Is Female to Male as Nature is to Culture?', in M. Z. Rosaldo and L. Lamphere (eds.), *Woman, Culture and Society*, Stanford University Press, Stanford, Calif., pp. 67–68.
46. Popper, K.: 1959, *The Logic of Scientific Discovery*, Basic Books, New York.
47. Ravetz, J. R.: 1971, *Scientific Knowledge and Its Social Problems*, Oxford University Press, New York.
48. Rich, A.: 1976, *Of Woman Born, Motherhood as Experience and Institution*, W. W. Norton & Co., New York.
49. Rorty, R.: 1981, *Philosophy and the Mirror of Nature*, Princeton University Press, Princeton, N.J.
50. Rose, H. and Rose, S.: 1976, *Ideology of/in the Natural Sciences*, Schenkman, Cambridge, Mass.
51. Rosenberg, R.: 1983, *Beyond Separate Spheres: Intellectual Roots of Modern Feminism*, Yale University Press, New Haven, Conn.
52. Rossi, A.: 1977, 'A Biosocial Perspective on Parenting', *Daedalus* **106**, 1–32.
53. Rossiter, M.: 1982, *Women Scientists in America: Struggles and Strategies to 1940*, Johns Hopkins University Press, Baltimore, Maryland.
54. Rowell, T.: 1972, *The Social Behavior of Monkeys*, Penguin Books, Harmondsworth, England.
55. Sayers, J.: 1982, *Biological Politics: Feminist and Anti-Feminist Perspectives*, Tavistock Publications, London.
56. *Signs: Journal of Women in Culture and Society:* 1978, special issue entitled 'Women, Science and Society', **4**(1).
57. Sohn-Rethel, A.: 1978, *Intellectual and Manual Labour*, Macmillan Publishing Co., London.
58. Stacey, J.: 1983, 'The New Conservative Feminism', *Feminist Studies* **9**(3), 559–84.
59. Stehelin, L.: 1976, 'Sciences, Women and Ideology', in H. Rose and S. Rose (eds.), *Ideology of/in the Natural Sciences: The Radicalization of Science*, Macmillan, London, pp. 76–89.

60. Stoller, R. J.: 1974, 'Facts and Fancies: an Examination of Freud's Concept of Bisexuality', in J. Strouse (ed.), *Women and Analysis: Dialogues on Psychoanalytic Views of Femininity*, Grossman Publishers, New York, pp. 343–64.
61. Tiger, L.: 1970, *Men in Groups*, Vintage Books, N.Y.
62. Tobach, E. and Rosoff, B. (eds.): 1978, *Genes and Gender I: On Hereditarianism and Women*, Gordian Press, New York.
63. Tobach, E. and Rosoff, B. (eds.): 1979, *Genes and Gender II: Pitfalls in Research on Sex and Gender*, Gordian Press, New York.
64. Wilson, E. O.: 1975, *Sociobiology: The New Synthesis*, Harvard University Press, Cambridge, Mass.

ROBERT C. SOLOMON

HETEROSEX

> Doggie style is the real thing. . . . . rear entry is the configuration that summons up atavistic memories of rutting in cold caves. . . . .
>
> Yet this is a new age. Maybe the day of hot tropical sex groundhog fashion will fade, and doggie style will cease to be its quintessential kind. A prediction for the near future? I'd bet that within a decade or two, the real thing among sexual positions will take a highly novel form: BOTH PARTNERS ON THE BOTTOM. "Impossible!" Inconvenient or uncomfortable, perhaps, but never impossible. (If they can land men on the moon, etc.)
>
> ([1], pp. 112–113)

In the excitement, confusion, liberation, oral and moral indignation inspired by the rediscovery of homosexual love and legitimacy, heterosex has been somewhat left behind. Adam and Eve have become a tired paradigm, a fashion too established and too familiar to inspire philosophical speculation and excitement. Heterosex is – what could be worse? – *normal* sex, subject to problems and consequently cures, of course, but nevertheless the standard which, however embellished or even lusty, beckons from the pages of *Playboy* with the smug satisfaction of an old conservatism.

In the solemnity of the current sexual counter-revolution, heterosex still does not get its due. The new disease is 'inhibited sexual desire' (ISD). (A decade ago it was the failure to have an orgasm.) The one-night stand is out; caution and commitment are in. But what gets lost in this newly fashionable conservatism is the 'joy of sex' itself. Sex has become once again a mere 'function,' at best 'an expression of intimacy' [9]. It is once again a mere 'desire,' which, typically, loses out to careers in college [5].

Whatever heterosex is or could be, it deserves something better than

E. E. Shelp (ed.), Sexuality and Medicine, Vol. I, 205–224.

its role as the medical measure of normal 'sexual functioning,' the possibly passionless but in the standard case modestly emotional ejaculation of male into female, in the appropriate place, in the appropriate amount of time, with appropriate expressions of satisfaction on both sides, during or immediately following 'intercourse.' D. H. Lawrence may have gone in a bit over his head, so to speak, but at least he was clear in his blasphemous piety about the unappreciated significance, the wonder of heterosex. Whatever the political or emotional virtues of homosexuality, heterosex embodies a union of differences and confusions which quite properly keeps it center stage. Indeed, it continues to provide the prototype for every alternative form of sexuality, from bestial to masturbatory. Heterosex is – sex.

What is sex? It is, in some sense, any coupling of warm bodies in certain not so strictly circumscribed ways. Some would say that it includes shaking hands, but we can probably restrict our attention to the bulkier body entanglements. The central paradigm, which provides us with the 'natural' continuity with most of the animal kingdom and even a small minority of plants, is heterosexual intercourse. Indeed, from the flat and untitillating use of the word 'sex' in biology textbooks to describe the behavior of scorpions and fish, not to mention pine trees, one gets the idea that sex for us is much the same physical process, except that we alone in the kingdoms of life have chosen to make a moral issue of it. The cynical conclusion is that sex is 'no big deal,' or the conspiratorial suspicion (that we find in Philip Slater, for example) that the most readily available commodity in the world has been made artificially scarce as a means of political control. But this is wholly misleading, if it is not actually fraudulent. In fact, most human sex has about as much continuity with arachnids and dicotyledons as eating the Sunday wafer and sipping the Passover wine have to do with nutrition. The medical measure of normalcy is as misplaced as the school nutritionist who oversees communion. Sex, like salvation, is spiritual. It transcends (though cannot ignore) the body.

Sex is defined by our ideas, not our bodies. Bodies to be sure are the vehicles of meaning, the media through which those ideas take form. But our conceptions about sex are far more tied to our morals than to our gonads, though, to be sure, one can easily and amusingly imagine how different our sexual fantasies might be if our organs were differently constituted, say, on the bottom of one foot or – reminiscent of a once popular but physiologically dubious movie – inside of the mouth or the throat. Even more imaginatively, we can speculate on the delightful or horrifying consequences of the introduction of a third sex, perhaps even so designed so that the so-called 'sex act' would not be complete without all three sexes. But we are limited – for better or worse – to two. This essay, accordingly, is about heterosex and its conceptions, and the rich and varied meanings that inevitably escape the overly medicinal and functional perspective in which our most personal passions have been confined.

THE TELEOLOGY OF SEX: 'THE PURPOSE OF SEX'

Sex is not just 'matter in motion' (we may refrain from mentioning the many more imaginative metaphors), plus the pleasurable sensations thereby promoted and produced. Sex, like virtually all human activities and like most biological phenomena, has a purpose, or purposes. In the perspective of biology, of course, that purpose is clear enough; the perpetuation of the species. Pleasure, the expression or the stimulation of romantic love, getting even with the husband or getting a good grade the hard way are at most epiphenomena, perhaps effective motivation but probably of no real relevance at all. Yet only rarely, perhaps never, does one 'have sex' within the biological perspective. The purpose of sex is therefore only rarely, if ever, the same as 'nature's purpose' – the perpetuation of the species. There is also *our* purpose or purposes, what we desire over and above – or instead of – fulfilling our biological roles as self-perpetuating links in a chain that extends from Adam and Eve or perhaps rather Lucy to Nietzsche's *Übermensch* or 'the last man.'

Philosophers – following the ancient Greeks – like to refer to the purpose of a phenomenon as its 'telos'; thus we can speak of the 'teleology of sex' – its ultimate purpose. But our teleology immediately splits at least in two, for there is what we have called, with evolutionary naiveté, 'nature's purpose,' and then there are *our* various purposes, which are not the same thing. Indeed, once we introduce that distinction, we immediately begin to wonder how the two have ever been connected, except for the contingencies of biology – in the same way that one might wonder how it is that the miracle we call 'language' has such an intimate connection with the organ of ingestion (though not, perhaps, for Jacques Derrida, for whom language is essentially 'writing' and self-presence is a problem).

'Nature's purpose'

To talk of 'nature's purpose' is, of course, naive. Nevertheless, it might be true. We no longer talk so easily of the teleology of the world – as Aristotle did so long ago and as Hegel and his henchmen did with less ease a century and a half ago. We prefer causal models, efficient rather than final causes, physiochemical processes rather than instincts and collectively unconscious cunning. Nevertheless, there is a sense in which we can say, uncontroversially but a bit oddly, that sex does *serve* a purpose – whether or not we want to say that sex *has* a purpose. Sex provides – for the moment – the only means of reproduction of human beings, though whether this is a good thing in the eyes of some larger telos is an awesome question I would not want to broach here. And sometimes, about 2.3 periods per lifetime for most Americans (considerably less for many couples in other cultures whose birth rate is inversely higher), nature's purpose does indeed coincide with at least one of the purposes of the heterosexual couple having intercourse. (It is logically possible, of course, that a homosexual couple might also adopt this telos, but this possibility is not a fruitful topic of discussion.) This happy harmony is not our concern here, however; it is the divorce rather than the marriage of sex and conception that interests us. Let us turn, therefore, to persons' purposes, leaving nature to fend for herself.

Persons' Purposes

It is with persons' purposes that the teleology of sex becomes complicated. One might mention some parallel complications in nature, for example, male dogs or chimps mounting other males, not in procreative confusion but as an unchallengeable gesture of dominance. I do want to consider such behavior, however, in the context of human sexuality. To take a more common example, it is often said, in many languages, that to have intercourse with a woman is to 'possess' her. I do not take such talk lightly, and the fact that the woman in question may or may not conceive as the result of the possessive act does not seem to be relevant to the matter. I choose this somewhat feudal example to underscore the sometimes less than romantic theme of the theory that I will be presenting to you here; sex is not, in addition to its biological functions, just a matter of pleasure or the expression of love. The new talk about sex as an expression of love and intimacy is just as naive as the sexual revolution's severing of them. The human purposes of sex are spread across an enormous range of symbolic and practical concerns, from love and intentional debasement to a mere sleep aid. To talk about 'the purpose' of sex, even confining ourselves to 'our purposes' is a serious mistake. It does not follow, however, that this diversity makes impossible a unified theory of sexuality. Indeed, it is the diversity itself that has to be explained.

When I talk about 'persons' purposes,' I do not mean *conscious* purposes, nor, of course, do I refer only to explicit and mutually agreed upon purposes. The cooperative telos of the transparently self-reflective and psychologically explicit sado-masochist couple is one thing; the fumbling, confused clash of desires, fantasies, habits, expectations, and aims that most of us take to bed with us is quite different as well as more common. Indeed, it is part of the ethics of sex (as opposed to the ethics of reproduction) that most of our purposes remain unspoken; it is a matter of curiosity as well as 'liberation' that we have recently been encouraged to make explicit our fantasies and desires. Not surprisingly, most of these tend to be rather mechanical 'touch me here' rather than the more significant messages and meanings that such behavior expresses, knowingly or not.

I do not want to attack the profound question of 'the unconscious,' much less the even more problematic Jungian notion of a 'collective' unconscious. In one sense, there is no doubt such a thing, consisting at least of those most basic biological impulses and residual instincts, what Kraft-Ebing misleadingly called 'the 6th genital sense.' But whether there is a *symbol*-laden unconscious is quite another matter, and at least one proto-Jungian philosopher, Nietzsche, vehemently denied the collectivity of our impulses, turning the tables to insist that our 'herd-consciousness' is to be found in consciousness, our individuality in our instincts. Our own view of sexuality, as opposed to some of our theories about it, seems to side with Nietzsche. We seem to belabor the illusion – against all evidence – that each of us is sexually unique. But – back to the point – many of our purposes, individual or collective, instinctual or learned, are not conscious. There can be no doubt that there are a number of inborn instincts involved in our sexual desires, but it is not all clear that there is any way of even beginning to say what these might be. Comparisons with less inhibited creatures may be amusing, but our own instincts are so plastic and/or overladen by learning that 'what we really want' – in some primitive way of speaking – must remain forever unknown to us.

"WHAT DO I WANT WHEN I WANT YOU?"

> my body – in which I am lodged like an ice cube in a furnace . . .
>
> ([4], p. 89).

Any discussion of human sexuality involves more than a catalog of organs, feelings, physiological and behavioral responses; sex is first of all desire, 'first of all' both temporally and, less obviously, phenomenologically. Indeed, one might make a case that sex without desire is not sexuality at all, though one hesitates to suggest or imagine what it might be. But what is it that one desires?

The glib answers are soon forthcoming, along with that familiar smirk and a number of rude responses. Such responses, however, are largely limited to adolescents and academic seminars. A novice well may think that the 'end' (telos as well as terminus) of sex is successful intercourse (though even the notion of 'success' should be carefully scrutinized here). But it is clear that, despite certain philosophical protests, sex is not one of those activities that is 'desired for its own sake' (if anything is so desired). It is not just that sex *sometimes* serves a telos, but that it *always* does. If there is anything like a purely sexual impulse (and they are very rare if they exist in us at all), we would still have to ask what it is an impulse to *do* – and do *to whom*. If it is 'the satisfaction of the impulse' that is in question, it is not obvious that sex is what satisfies it. As Freud often pointed out, there can be 'physiological' satisfaction without 'psychic' satisfaction, and it is by no means clear (especially to Freud) whether the sexual impulse is primarily 'physiological' or 'psychic,' which was discovered or 'postulated' – in fact fabricated – at the end of the 19th century. Some animals, perhaps, may experience something like 'pure sexual desire,' but this would be noteworthy only insofar as it is also isolated and empty, virtually unrelated to any other activities and something of a mystery to the animal itself. (Dan Dennett, in an unusual bit of anthropomorphizing, imagines a bird in the midst of its instinctual behavior, musing to itself, 'why am I doing this?') It is clear that few mammals would enjoy such an impulse; sex becomes a matter of status as well as species-preservation, and, watching my male dog's behavior around a female in heat, it becomes quite clear that sexual desire supplies only a small if not minimal motive, and that interwoven inseparably with a half dozen other concerns. And when we return to human sexuality, we might be far closer to the mark, not with the obvious but with the obscure, not with the simple-minded aim of wanting intercourse but with the 'infinite yearning' of *eros* suggested by Aristophanes in Plato's *Symposium*. In philosophy, anyway, it is always better to err towards the infinite than to get trapped in the picayune.

So, what is sexual desire? Indeed, is it possible (as Michel Foucault sort of suggests) that we ought to give up such talk as quite misleading and mythological, as if sex were indeed a distinct and isolated impulse with only the most peculiar and often antagonistic connections with our other motives and desires? But without going quite so far, we can begin to appreciate the complexity of what we call 'sexual desire' and its intricate connections with other desires and motives. Sex is not a distinct activity except according to certain customs and rituals that make it so. Accordingly, sexual desire is not an easily distinguished mode of desire – which is how it is possible to produce those yearly disclosures that football, Pac-Man, quoits, metaphysics, process philosophy, and eating artichokes are *really* sexual. Who can say where one desire starts and another ends. How is it that Eric Rohmer's obsessive desire to touch Claire's knee is highly sexual, while the desires of a jaded Don Juan may be about as sexual as his desire to get it over with and go home to watch the Tonight Show? When we begin to analyse sexual desire, therefore, we should not be surprised that there is much more to it than sex. Indeed, sometimes sex might not even be part of it at all.

In somewhat Freudian terms, we might say that both the aim and the object of sexual desire is indeterminate. The end of sex need not be sexual intercourse, and it is not at all clear which aspects of a person are desired. (Both 'your body!' and 'the whole person, *all* of you' are unimaginative fictions). Claire's Knee is a playful bit of fetishism, but it represents well enough that body-part consciousness that provides the focus of many male desires. Jean-Paul Sartre used to say (to Simone) that he disliked the 'final act' of sex, but went through it only because he was expected to. Sex short of intercourse need not be frustrated, inhibited, or truncated. It may be all that there is. Indeed, one might be obsessed with sex but find the very idea of intercourse unthinkable – and one should not leap to the now nearly automatic conclusion that such a desire must therefore be 'repressed' or 'suppressed' or any other 'pressed' including 'un-ex-pressed.' But if sexual desire is not necessarily aimed at intercourse – heterosexual intercourse in particular – then the supposed essential tie between sex and sexual desire and those rather constricted activities in which pregnancy is possible is dramatically weakened. This is not to deny, of course, that most (heterosexual) sexual couplings do end (but as terminus, not telos) in intercourse of the most often prescribed variety. It is only to say that sexual desire – including the desire for intercourse – is not so firmly connected to sexual intercourse as is so often supposed. It is the *meaning* of sexual desire that escapes us, though by this I certainly do not mean to allude to the mysteries of love or the theological speculations concerning sex and God's will. One of the problems with sex – as with so much in our vulgarly Cartesian society – is the dualism between the 'merely material' on the one hand and the spiritual on the other, as if 'spirit' referred only to the transcendental verities of religion. But Hegel had it right: 'spirit' (*Geist*) refers to the whole of human intercourse. Accordingly, if sex isn't spiritual, it would be difficult to say what is.

FUNCTIONALISM AND THE MEANING OF SEX

> Here as elsewhere, the Doktor's thinking seems too narrowly phallic. The loss of the penis is but one of the countless forms of human powerlessness. We thrust not only with our penises. Perhaps the castrated member is itself but a symbol?
>
> ([4], p. 21)

In a society schooled in efficiency, in which ethics and utility are often thought to be the same, in which even physical exercise is supposed to serve a purpose, we should not be surprised that our conceptions of sex tend to turn to questions of function, to the *uses* of sex. My discussion of the teleology of sex might well be viewed as part and parcel of this pragmatist orientation, and desirable as it might seem 'for its own sake,' sex in America is almost always defended in terms of some further goal – good health, happiness, interpersonal satisfaction. The idea of true chastity horrifies us, and medical science has not been slow through the ages to catalog the various malfunctions that cruelly follow this unnatural abstinence.

But yet, the pragmatic notion of 'having a function,' like the more classical notion of teleology, is not a singular notion. It admits a variety of interpretations, from the traditional biological accounts of 'instinct' and 'species preservation' (that is, 'nature's purpose' – serving that natural function) to the Freudian account of pleasure as the purpose, construed, however, in terms of the hydraulics of the 'psychic apparatus' not the phenomenal sensations of 'fun.' Masturbation, accordingly, gets raised to the status of a full-fledged sexual activity, indeed, perhaps, the primary sexual activity. (It is worth noting that such a curious view was common in German intellectual circles long before Freud. Even Immanuel Kant, who never tried it, suggested that sexual intercourse was 'mutual masturbation,' over a century before Freud's speculations into 'primary processes.') However these interpretations differ, what they have in common is their functionalism – that is, their emphasis on some physical or quasi-physical function of sex, devoid (in its essence at least) of *meaning*. This too is the difference between functionalism and teleology, properly speaking: functionalism refers only to function, while a *telos* tells us what an act *means*. It is worth noting how easily some discussions of sex's natural function as reproduction turn into discussions of 'procreation,' which in a word takes us an infinite step beyond mere function. So too 'fun' is often a *telos* disguised as a function. The pleasures of sex are not just sexual pleasures.

To say that sex has a telos is not the same as to say that sex serves a 'function,' whether this be the preservation of the species or the release of psycho-physical energy. Functionalism, crudely conceived, is the view that sex serves the conveniences of life (which is not to say that sex is always convenient). Sex is not an end in itself, but the realm of the sexual is something more than biology and appetite, more philosophical than functional. Again, let's remember Aristophanes' reference to *eros* as an 'infinite yearning.' We will not understand that yearning unless we give up the functional as well as the limited biological view of sex. The genealogy of 'sex' begins, no doubt, with the birds and the bees. But it is a thin 'natural' history of love that does not soon see that the very word 'natural' – in cultural context – is a *moral* term. 'Nature' might aim at propagation of the genotype, but cultures are rather concerned with group solidarity, rank and status, rules, respect, and, for the sake of authority as well as order, prohibitions. To think that all of this is functional in the crude sense, aimed at population or emotional control, is to expand the scope of functionalism beyond intelligibility. It is rather teleology – concerned with human goals and competitions, cultures and conventions, our most intimate means of domination and submission, humiliation and possessiveness. The history of sex is the history of power politics. What is 'natural' in sex depends not on biology but on ideology.

It is often said – particularly by those attacking homosexuality – that heterosex is 'natural.' But the concept of 'nature' is our inheritance from Aristotle and the Greeks, and then from the church, then the Enlightenment. In all of its forms, it displays one of the more glamorous sophistries in Western thinking, the appeal to a silent and therefore incontrovertible Nature in the defense of particularly provincial practices. Indeed, it is rarely biological nature that is in question; it is more usually 'natural reason,' which is, it has often been suggested, most unnatural, even 'a bit of the divine.' When philosophers sought a 'natural religion,' for example, they were not after nature worship or what Santayana called 'animal faith,' they were after rational arguments. And when there have been demands for 'natural sex,' it is not bestiality or rear-entry intercourse that is preferred; the call is rather to 'rational sex,' which means that for which there are accepted standards and arguments, most notably what I have elsewhere called the male-minded two-minute emissionary missionary male-superior ejaculation service (with some credit to John Barth). Whether or not sex is the danger to civilization that Freud thought it was, the traditional emphasis has indeed been on quick and efficient sex, just enough to satisfy what D. H. Lawrence has called 'evacuation lust' and, in the larger scheme of things, insure an adequate supply of progeny to maintain the status quo. Indeed, in our genealogy of sex and 'sex,' we should look to the social politics of classes and inheritance rights as well as the seemingly natural passion of lust.

It is in this political light that we should view the seeming 'naturalness' of heterosex. We should not forget that our most celebrated example of culture – the ancient Greeks – were homosexuals who despised women and heterosexuality and considered what we would call 'romantic love' most unnatural. In Plato's *Symposium* virtually all of the speakers take pains to point out the inferiority of that merely functional domestic *eros* in contrast to the divine status of that 'higher' *eros* between men (more properly, between youths and men). But do we find when we look at Greek homosexuality a licentious free-for-all with no restrictions, given what we would call the 'unnaturalness' of their passions? Not at all. Indeed, the best books on the subject (cf. [3]) as well as the ancient texts themselves show us quite clearly that the sexual ethic among the Greeks was, if anything, more precise and much less hypocritical than the sexual mores of modern heterosexual cultures. What was 'natural' was in fact a matter of convention, not biology. (The Marquis de Sade once commented, while railing against the popular 'argument from design' that was circulating in 'natural theology': if God had not intended the anus to be used for sex he would not have shaped and positioned it so conveniently.)

WHAT IS HETEROSEX - SEX, GENDER OR ORIENTATION

> The claw (of the fiddler crab) is waved in a ritualistic way so as to provide a gesture absolutely characteristic of the species so that there can be no mistake in identity . . . In all cases, since such movements are mostly communication mechanisms, the visual apparatus and perceptual abilities must be equally exactly adjusted to recognize, and distinguish from all other visual stimuli to which the animal may be exposed, the significant movements which indicate the right sex and the right species in the right phase of the breeding cycle.
>
> ([11], pp. 143–144)

No aspect of heterosex is so misunderstood – or so taken for granted – as the seemingly plain fact that there are two sexes. 'Boys will be boys and girls will be girls' has proven to be an undependable assumption in education and politics. It turns out to be similarly untrustworthy in sex. The fiddler crab may be blessed with an unambiguous 'hard-wired' circuit for sexual recognition and performance. We are not. Indeed, in certain parts of the urban ecology today the mere recognition of males and females is not the easiest part of it.

Heterosex is not just sex between two members of 'opposite' biological sexes. It is also a set of cultural conventions, a way of upbringing, a rich set of mythologies, not least of which is the myth that 'boys will be boys, etc.' How one perceives others sexually depends in part on how one conceives of oneself, and 'I am male' is at best a crude beginning. (Compare, for example, the rather dramatic differences between 'I am a female' and 'I am a woman.') It is essential to distinguish, therefore, between at least four different dimensions of heterosexual identity:

(1) biological sex
(2) gender identity
(3) sexual orientation
(4) sexual ethics

Biological sex refers to 'male or female,' but this is not quite so clear as one might suppose. John Money, director of the John Hopkins Gender Identity Clinic, has shown that individuals are often 'intersexed' and possess either internal or external organs of the opposite sex. The much discussed complications of X and Y chromosome duplications mark only the origins of such complications. Fetal development, hormonal changes, and brain differences are responsible for a broad spectrum of biological sexual differences. Indeed, Money has distinguished at least five distinct categories, depending on (1) chromosomal configuration, (2) gonadal sex, (3) hormonal sex, (4) internal reproductive structure, and (5) external genitalia. It is clear that most of heterosex relies quite simply on criterion #5, as if the rest of one's sex can simply be taken for granted.

Gender identity is typically considered to be nothing but the extension of biological sex. It is not. What we call 'masculinity' and 'femininity' are now generally recognized to be learned, cultural roles. Money has shown that children who are 'assigned' a sex contrary to their chromosomal sex grow up with a sense of gender appropriate to their assignment, not their genes. Cultural expectations lead in turn to differences in social treatment that further reinforce the sense of gender identity (cf. [7]). Gender identity permeates one's sense of self, and it consequently colors most of our interpersonal relationships as well. Heterosex is not just a kind of sex; it is – for better or worse – one of the structural categories of our society. And, to make our post-feminist lives more interesting, we might note that this has only a contingent relationship with the facts of sexual identity.

Sexual orientation is distinct from gender identity in a number of ways. It is clear that asymmetrical concepts of 'masculinity' and 'femininity' define many homosexual and lesbian relationships; it is not so obvious but just as evident that the gender identities that make up heterosexual relationships may be just as complex and varied. We are, in this essay, only concerned with heterosex, but the variation in sexual orientation – the choice of sex 'objects' – may be spectacular even within this framework. Much of what is relegated to the 'subjective' realm of 'personal attraction' is in fact the much more profound question of orientation, in which one's whole concept of sex and self comes into focus in some very hard to recognize ways. Consider the aggressive woman who always finds herself a 'mouse,' a man who always seems to need a woman who is wholly dependent on him. Orientation is not just a matter of personal 'taste.' It cuts to the quick of the concept of heterosex as such.

Sexual ethics is something more than the morals of marriage and mutual respect. Indeed, as we all know, something rather odd and sometimes disturbing happens to the ordinary rules of morality when one enters into an intimate relationship. Heterosexual sado-masochism is the most dramatic example of a case of 'consenting adults' that seems to us both unacceptable and 'none of our business.' But virtually all of sexuality falls into the same dilemma: our ordinary notions of 'privacy' break down in a dramatic way. The meaning of 'respect' alters considerably. New obligations come into being as old formalities melt away. How we treat each other, in other words, itself becomes part of our sexual identities, defines those identities, and, with them, defines heterosex as well.

SEXUAL PARADIGMS

To say that sex is teleological, to insist that sexual identity has much to do with sexual ethics, is to assert something more than the now familiar thesis that sex is largely cultural and conventional. It is to insist that our sex depends on our philosophy. What sex *means* depends on what we take it to mean. Even sexual performance and pleasure depend on the ideas with which one enters into sex. To say the obvious – believing that sex is 'dirty' inevitably makes sexual experience unpleasant – seemingly intrinsically so. To conceive of sex as power or as love is to define sexual experience in a certain way. Sex 'in itself' is neither natural nor functional. Our ideas determine what is natural. Our philosophies give significance to functions. I have suggested that heterosex serves as a paradigm of sex, but this, now, appears to be too simple. In fact, heterosexual intercourse is not itself a paradigm so much as it is the convergence of many paradigms, and the job of a philosopher – in addition to enjoying the practice – is to develop a theory of heterosex in which these various paradigms are distinguished and played off against one another. In this final portion of this essay, that is what I should like to do briefly.

I would like to describe four distinct sexual paradigms. There are more, but my aim is just to familiarize the reader with the very idea of a sexual paradigm, so jaded we are with the idea that sex is just obvious. Indeed, it is not obvious at all, and I might add that our sexual paradigms are one of the cornerstones of our conceptions of ourselves; and analysis of sexuality is therefore – as Sartre somewhat brutally pointed out – at the very core of an understanding of our being-with other people.

The four paradigms are:

(1) the reproductive or, more piously, the procreative
(2) the pleasure or 'recreational' paradigm
(3) the metaphysical or 'Platonic' paradigm
(4) the intersubjective or communication paradigm

The Reproductive Paradigm

The reproductive model has perhaps always been overemphasized, but, in the religious dress of 'procreation,' it has been especially well-nourished under the auspices of the Christian church. It is a paradigm that distills off the pleasures and romantic overtones of sex and reduces it to a matter of dutiful service to the species or to God. Of course the paradigm is typically embellished and not left in its cold biological state, but even so it is distinctively unconcerned with questions of mutual expression and tenderness. (The union of souls as well as bodies is a secondary development.) The paradigm of reproductive sexuality is the effective and efficient ejaculation of the male into the female. His orgasm is a bonus; hers is of no importance whatever. Indeed, excess enjoyment, wasting time and – most essentially – 'wasting seed,' are perversion. Indeed, although the concept of 'perversion' did not really enter into sexual ethics and medicine until the 19th century, the prohibitions against masturbation, sodomy, and all other manners of 'wasting seed' are to be found clearly in even the Old Testament. (What must the Athenians have thought of that?!) Needless to say, the sole paradigm example here is heterosexual intercourse. Homosexuality has no place whatever, and even love and pleasure are of at most secondary importance, as enticements, not telos.

Though nourished by religion, the reproductive model does not at all depend on religion for its persistence. Kant (who was also pious) insisted that our sexual appetites are justified *only* insofar as they serve Nature's End, not our own. So too, politics or simple excessive biological awareness can promote the reproductive paradigm. A Darwinian might well see sex this way; a couple desperate to have children may temporarily conceive of their sexuality in this way. A sexual moralist might literally refer to 'natural sex' – in an argument against homosexuality, for example – without invoking religion at any point. Indeed, this is often the more powerful and harder to refute version of moral arguments concerning sexuality, that our 'natural' desires are aimed at reproduction, whether or not *we* aim at reproduction – or even thwart it – in our sexual activities. One might suggest, for instance, that such sex in general is 'practice' for the real thing, an odd suggestion that has, nevertheless, a considerable history behind it.

A paradigm is not necessarily an exclusive perspective, a set of blinders permitting only a single telos or conception. It is rather a matter of *focus*, a way of picking out certain features as essential and pushing others into the background. The reproductive paradigm highlights the biological consequences and minimizes the recreational and expressive features of sex. The procreational version of the paradigm adds to this the theology of divine design. Neither version need eliminate or ignore the other aspects of sexuality, but their emphasis already establishes an ethical framework. It is this framework within which the morals of contraception become heated issues, and it is within this framework too that the technology of contraception has the greatest effect on our sexual conceptions. But there are other paradigms, and with them, other ethics and other concerns.

The Pleasure Paradigm

The pleasure paradigm might be called 'liberal' as a reaction against the more conservative reproductive/procreative paradigm. It is, for the most part, the paradigm that provided the ideology of the 'sexual revolution' of the '60s, but contrary to certain neo-Freudian hysteria, the '60s were not the time of '*eros* unchained,' as one popular book of the times would have it. There was not a massive libido suddenly released. Rather, there was a shift in paradigms, a swing of the *Zeitgeist* that included, not coincidentally, a new attitude toward the meaning of sex.

The real father of the sexual revolution was Sigmund Freud. It was Freud, for all of his personal sexual conservativism, who most soundly attacked the traditional reproductive/procreational paradigm and put in its place a model of sex with pleasure as its telos and its defining characteristic. Sex, for Freud, is not Nature's purpose but rather the primary drive of the 'psychic apparatus.' And yet, Freud's pleasure paradigm, like the reproductive paradigm, rests on a biological foundation. It is the nervous system that is the locus if not the focus of Freud's theory, and its physiological basis that determines the telos of sex. The goal of the nervous system, and of the organism, is 'discharge,' Freud tells us. Its principle is homeostasis, or 'the constancy principle.' Its ultimate aim is emptiness, relaxation (a formulation that later led Freud into his flirtation with 'the Nirvana principle' and 'the death instinct'). The release of built-up tension is pleasure; the retention of tension is pain. The release of 'catharsis' of tension in sexuality produces the orgasm, and it is the pleasure of the orgasm – that is, the pleasure of release – that is the end of sexuality, not the procreative consequences of ejaculation. Thus it is important to note how central Freud's distinction between physical and psychic satisfaction becomes for him; it is the latter that is essential to successful sex, not the former. This simple hydraulic picture is complicated by the fact that in addition to this 'primary process' there is a complex system of Ego-needs, which involve identifications with certain sexual 'objects.' Accordingly, sex becomes concerned with the satisfaction of these secondary drives as well, and the simple pleasure of sex is complicated if not thwarted by the need for acceptance, security, and love. One might note the echo of Jean-Jacques Rousseau here, for Freud was also a 'romantic' of sorts, if also a pessimist *à la* Schopenhauer rather than a hopeful visionary like Rousseau. One might also here take full note of the fact that, as release of tension, masturbation plays a central role in Freud's notion of sex, while on the reproductive paradigm this is at best a perversion, if not irrelevant to sex altogether.

The notion of 'discharge' or 'catharsis' in Freud's paradigm is obviously based on a male-oriented perspective. In this as well as in its biological foundation, it is not entirely distinct from the reproductive model. Understanding Freud, however, is not primarily a matter of the history of medicine so much as it is a socio-cultural concern, and his rebellion against the reproductive model was part and parcel of a far more general rebellion against a certain kind of society. So too, the pleasure model in its more modern version, whether the ACLU version that allows any sexual activity between 'mutually consenting adults' or the more vulgar version that simply insists, 'if it feels good, do it,' consists in part of a rebellion against restrictions no longer acceptable, paradigms no longer unchallengeable. As a 'liberating' paradigm, the pleasure model needs no defense. A more difficult question is whether it is the best paradigm, a question that has become increasingly urgent in the counter-revolutionary years of the last decade.

What ails the pleasure paradigm? Fifteen years ago Rollo May wrote in a best-selling book, *Love and Will*, that the sexual revolution has wrought its own forms of repression and unhappiness. His argument turned too much – in accord with that psychiatric genre – on the pathological, on the failures and victims of the revolution rather than its beneficiaries. Nevertheless, as both Freud and May have insisted with some justification, the plight of the neurotic may be a mirror – if a distorting mirror – of the pathology of everyday life. If the reproductive model tends to make sex too restrictive, the pleasure model tends to make it too conceptually permissive. Two hints from Freud: any model of sexuality that makes masturbation as central or more central than intercourse, any model that refers to the other person in sexual desire as an 'object,' is an unlikely candidate for an adequate model of most of our sexual lives. Again, the paradigm may be liberating, but that does not mean that it is wholly adequate in itself.

A proper evaluation of the paradigm would be an essay in itself, but let me mention just three main points of contention here:

(a) Aristotle argued 2500 years ago that pleasure is never the end of *any* of our activities, even including sex. It is rather the *accompaniment* of good activity, the enjoyment that comes from being virtuous and fulfilling one's telos. Thus the question becomes, what is the telos of sex such that it gives us so much pleasure? What is it that we satisfy? What do I want when I want you such that I so enjoy having you when I get you? To say that pleasure is the telos of sex is to confuse the play with its musical accompaniment. (In that context, a consideration of the very different moods of making love, and the music that one might play along with them, is not unrevealing.)

(b) Sex is mutual activity; masturbation, no matter how satisfying, is not our paradigm. At least part of the answer to (a) above is that there is satisfaction in being with or 'having' the other person, not as a 'secondary' concern, as Freud would have it, but as the primary concern. The promiscuity and lack of discrimination of horny males aside, it is not an 'object' of sex we seek, but a partner.

(c) Moreover, because sex depends on concepts and paradigms as well as basic biology, our sexual behavior is never 'without meaning.' Some meanings, however, are demeaning, and every sexual act has its significance. To think that one can indulge in the traditionally most powerful symbolic activity in almost every culture without its meaning anything is an extravagant bit of self-deception, which, nonetheless, has not prevented its wide-spread appearance in libertines and rakes of all ages.

The Metaphysical Paradigm

The metaphysical paradigm is by far the most poetic of the four, but I want to be brief and prosaic about it here. We demand 'meaningful' relationships; this paradigm provides meaning. The metaphysical paradigm is romantic love or *eros*; its classic test is Plato's *Symposium* (thus I would also call it 'the Platonic paradigm,' which is not to be conflated with the asexual notion of 'Platonic love' that emerged 15 centuries later.) In simple-minded parlance, it is the vision that two people were 'made for each other.' Protestant Christianity often supplements or even replaces the procreative paradigm with this metaphysical paradigm, thus shifting the religious as well as the sexual emphasis to love (*agape* not *eros*). The best picture of the metaphysical paradigm, however, is Aristophanes' wonderful tale in the *Symposium* of a double creature cleft in two by Zeus ('like an apple') and ever since trying to find its 'other half.' And indeed sex does sometimes seem like that, an experience so overwhelming and filled with significance that questions about the biology of sex simply seem vulgar and quite beside the point.

The Intersubjective Paradigm

The intersubjective or communication paradigm is, as one might guess, the paradigm I would most like to defend, but I will not try to do so here. What I would like to do is offer it through an example, and then use it to conclude why it is that, in the course of my research, I have decided to minimize the seemingly obvious influence of sexual technology on sexual attitudes and concepts. The example comes from Jean-Paul Sartre, in his great 1943 work *l'Être et le néant* (*Being and Nothingness*) [8]. For Sartre, unlike Freud, sex is *essentially* interpersonal. He does not deny that masturbation is sexual; there just isn't anything very interesting to say about it. It is not just the consequences of sex that count; it is its *meaning*. But where the metaphysical paradigm tends to take a rather rosy view of meaning, Sartre sees quite clearly that meanings can be otherwise. His dark vision is summarized in his famous play, *Huit Clos* (*No Exit*): 'Hell is other people.' Sex, accordingly, is essential *conflict*, or rather, it is the central battleground in our interpersonal wars. Sex is *intersubjective*; it is primarily concerned not with pregnancy or pleasure nor even togetherness, but rather with effecting as definitively as possible the other person's constitution of both him or herself and oneself as well. It is a battle of domination and freedom, Sartre tells. Why does sex so focus on the most inert parts of the body – breasts and buttocks? Not for pleasure and not for the sake of reproduction but, symbolically, Sartre says, 'to turn the other into an object,' to *reduce* him or her to a mere body in one's control. Every caress is manipulation; pleasure is an instrument, and one's own pleasure – far from being the goal of sex – may easily be a distraction and a defeat.

Now, I do not endorse this gloomy view which, if taken whole, would put most of us in the land of the voluntary chaste. But its structure I consider nothing less than revolutionary – more revolutionary than even Freud's pleasure paradigm and the much touted sexual revolutions of the past. What Sartre does for us is to shift the paradigm once again, to sex as *expression*. It is not one's own pleasure that counts primarily (though one need not be indifferent to it); nor is it even the pleasure one produces in one's partner. Mutual pleasure production is not the telos of sex, even when as so often it is the most explicit focus of sex. (Thus Kant, who died a virgin, once described sex as 'mutual masturbation.') One can correct Sartre's pessimistic picture by noting that love, too, is expressed through sex, but it is essential not to leap too quickly to the naive romanticism that says that *only* love is, or ought to be, expressed by sex. Heterosex is first of all a kind of poetry, and there is no clear limit to richness of interpersonal feelings that it can express, given an adequate vocabulary and what we might call, tongue in cheek, sexual literacy.

CONCLUSION

Heterosex, long the subject of the best and worst but in any event passionate prose and poetry of Western literature, threatens today to become, perhaps, even boring. Freud was wrong, for many reasons, when he argued that sex was an inner 'drive,' powerful and irrepressible even when 'repressed,' active and exciting no matter what the social and cultural context. The fact is rather that sex like every other passion requires cultural nourishment, social support, not just medical attention and lectures to adolescents. Freud worried that sex will become simply civilized, without secrecy, without mystery, another scientific transparency without a hint of a shadow. What renders sex as communication something more than ordinary conversation is the darkness of what is revealed – the fears, the vulnerability, the lust, the joy. What makes sex so exciting is, still, the fact that every act of sex is an act of rebellion, every sexual grunt and gesture a breakthrough, even in the most 'permissive' society. If sex is part of medicine, doctors should understand themselves as something more than professional healers, something more than moralists, something more than mere scientists dealing with a touchy problem. If medicine seeks to understand heterosex, physicians should study poetry, not just physiology. Heterosex may not make the world go 'round, but neither is it just part of the plumbing.

University of Texas,
Austin, Texas,
U.S.A.

ACKNOWLEDGEMENT

Parts of this essay have been adapted from my 'Sex, Conception, and Conceptions of Sex', in [2].

BIBLIOGRAPHY

1. Anderson, K.: 1980, *The Real Thing*, Doubleday, New York.
2. Bondeson, W. *et al.*: in press, *The Contraceptive Ethos*, D. Reidel, Dordrecht, Holland.
3. Dover, K.: *Greek Homosexuality*, Duckworth, London.
4. Goldstein, R.: 1984, *The Mind-Body Problem*, Random House, New York.
5. Lee, J.: 1984, 'The Revolution is Over', *Time* (April 9), 74–83.
6. Money, J. and Ehrhardt, A.: 1972, *Man and Woman, Boy and Girl*, John Hopkins University Press, Baltimore.
7. Sanders, J.: 1974, 'Parental Sex Preference and Expectations of Gender-Appropriate Behavior', unpublished masters thesis, Brooklyn College.
8. Sartre, J.: 1956, *Being and Nothingness*, H. Barnes, trans., Philosophical Library, New York.
9. Scharff, D.: 1982, *The Sexual Relationship*, Routledge and Kegan Paul, Boston.
10. Slater, P.: 1971, *Pursuit of Loneliness*, Beacon Press, Boston.
11. Thorpe, W.: 1974, *Animal Nature and Human Nature*, Methuen, London.

ELI COLEMAN

BISEXUALITY: CHALLENGING OUR UNDERSTANDING OF HUMAN SEXUALITY AND SEXUAL ORIENTATION

> "The time has come, I think, when we must recognize bisexuality as a normal form of human behavior."
>
> Margaret Mead ([26], p. 29)

While society has struggled to understand and accept the existence and normality of homosexuality, the struggle to understand and accept bisexuality has been even more difficult. Society, it seems, would rather accept the myth that sexual orientation can be divided into two distinct categories: heterosexual and homosexual. As Klein [22] has indicated, some of the fundamental myths of society regarding sexual orientation are: a person is either straight or gay; there is no such entity as bisexuality; the bisexual is really homosexual; and bisexuality is only a transitional stage; and the bisexual is simply mixed up and can't make up his/her mind. Yet, even though Kinsey and his associates published their data over 30 years ago, these myths continue to be perpetuated. Kinsey, Pomeroy, and Martin [21] stated:

> The world is not divided into sheep and goats. Not all things are black nor all things white. It is a fundamental of taxonomy that nature rarely deals with discrete categories and tries to force facts into separated pigeon holes. The living world is a continuum in each and every one of its aspects. The sooner we learn this concerning human sexual behavior the sooner we shall reach a sounder understanding of the realities of sex ([21], p. 639).

Anthropologists such as Ford and Beach have also recognized the lack of exclusivity of sexual orientation in humans and the animal kingdom:

> When it is realized that 100% of the males in certain societies engage in homosexual as well as heterosexual alliances, and when it is understood that many men and women in our society are equally capable of relations with partners of the same or opposite sex, and finally, when it is recognized that this same situation exists in many species of subhuman primates, then it should be clear that one cannot classify homosexual tendencies and heterosexual tendencies as being mutually exclusive or even opposed to each other ([16], p. 236).

E. E. Shelp (ed.), Sexuality and Medicine, Vol. I, 225–242.

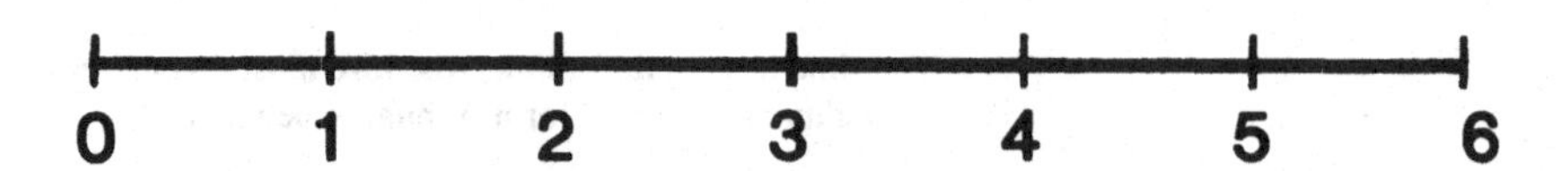

Fig. 1. Kinsey Continuum – 0 = exclusive heterosexuality; 6 = exclusive homosexuality

Consistent with this view of sexual orientation, Kinsey and his associates developed a 7-point scale to describe the continuum of sexual orientation (see Figure 1). Individuals are placed on this continuum based on their sexual behaviors and erotic attractions. Zero represents exclusive heterosexuality and six represents exclusive homosexuality. Three on this scale indicates equal same-sex and opposite-sex responsiveness.

In many ways, using this continuum is a far better classification system than the inflexible words 'homosexual' or 'heterosexual'. Even the word bisexual presents many problems because it can be used to describe so many different people and behaviors. Freud, for example, in his essays on sexuality [17] made distinctions of *amphigenic inverts* who were able to engage in satisfactory sexual behavior equally well between same and opposite sex individuals and *contingent inverts* who were able to derive pleasure from same-sex activity under certain circumstances or conditions. These types were seen by Freud as simply variants of heterosexuals and *absolute inverts* (exclusive homosexuals).

Gagnon [18] proposed five categories to describe different bisexual behaviors: young people experimenting with their sexuality; prostitutes who engage in same-sex activity for rewards; individuals who have ambivalent attachments to either men or women and these attachments are likely to be a transition from a heterosexual to a homosexual identity (or vice versa); individuals who are responsive to sexual stimuli regardless of its source; and individuals who have positive preference for both genders either simultaneously or sequentially based largely on emotional attachment criteria.

Klein [22] has offered yet another classification system of bisexual individuals:

Historical: one who lives a primarily heterosexual or homosexual life but has a history of bisexual experiences and/or fantasies.

Sequential: a person who has sexual relationships with men and women in a sequential fashion, with only one gender at a time.

Transitional: an individual who uses a bisexual orientation as a bridge to change his/her sexual orientation from one end of the continuum to the other.

Klein also recognizes that bisexual behavior may be episodic, temporary, experimental, or situational.

CAN A DEFINITION OF BISEXUALITY BE REACHED?

Different authors have attempted definitions that recognize the diversity of the phenomena. Some definitions that have been offered are as follows:

(A bisexual is) . . . one who is capable of a complex state of sexual relatedness characterized by sexual intimacy with both sexes [22].

The bisexual is a person endowed with both a homosexual and heterosexual option. Each option deserves equal social sanction, for the bisexual did not choose his duality, nor can he choose to eradicate it [27].

(Bisexuality is) . . . a well documented, normal human capacity to love members of both sexes [26].

To be bisexual means that a person can enjoy and engage in sexual activity with members of both sexes, or recognizes a desire to do so. Also, though the strength and direction of preference may be constant for some bisexuals, it may vary considerably for others with respect to time of life and specific partners [25].

In attempting to define bisexuality operationally, some have used Kinsey ratings. However, even Kinsey's continuum poses some dilemmas. Are bisexuals simply people who are rated '3' on the continuum? Or are they those who are rated '2–4' or '1–5'? In a survey of bisexuals published by Playboy Magazine [13], most individuals who self-identified as bisexual rated themselves as '1–3' on the Kinsey Scale. Similarly, Blumstein and Schwartz [4] found in their study of bisexuality that a number of men with little or no homosexual experience identified themselves as bisexual. These authors found little relationship between the amount of homosexual or heterosexual behavior and a person's choice to label themselves as bisexual, homosexual, or heterosexual.

These findings pose some serious dilemmas for the usage of '1–5' or any other combination to define bisexuality in any research population.

Some look at the label bisexual as a political statement just as many take on the label 'gay' or 'straight'. There are many individuals who label themselves as 'bisexual' or 'gay' as a way of affirming their bisexuality or their right to choose. Some bisexual individuals identify themselves as 'gay' which, for them, is a political and psychological attitude which acknowledges and affirms their ability to love and have sex with members of the same sex. Just as men can be 'feminists', individuals with little or no same sex activity can label themselves as 'bisexual' or 'gay'. Their label, therefore, may not describe their current sexual activities or relationships.

Another important point is to recognize that sexual orientation is not a permanent entity. In the Kinsey studies, respondents were rated for each period of their sexual histories. These multiple ratings revealed that sexual orientation fluctuates considerably over time. There is a myth that sexual orientation is constant. Major changes in sexual orientation may be more frequent than many believe [4, 5, 22, 25].

Next, I would like to clear up a conceptual confusion regarding sexual identity. Some do not recognize that there are four distinct aspects of sexual identity and that they are to some degree independent from one another. Shively and DeCecco [32] offered the following classification system to end this conceptual confusion. The four components of sexual identity are biological sex or physical identity, gender identity, social sex-role identity, and sexual orientation identity. *Physical* or *biological identity* is the biological and chromosomal identity of the individual. Bisexuality in this sense may refer to hermaphroditism, which can occur as a result of chromosomal malfunction or abnormality. But we should recognize as Freud [17] did that a certain amount of anatomical hermaphroditism or bisexuality occurs spontaneously. In every normal male or female, traces are found of the opposite sex organ structure. These traces exist in a rudimentary or modified fashion.

An individual's *gender identity* refers to the basic conviction of being male or female. This identity is a function of how one was reared rather than biological variables [28]. While physical identity has been established in the fetus and prior to birth, one's gender identity may not form until the age of two or three. We can see many clinical examples of continued gender confusion and bisexual gender identification in working with transsexual and transvestite patients.

One's *social sex role identity* refers to the identification with characteristics of individuals which are culturally associated with one or the other sex. Shively, Rudolph and DeCecco [33] have identified six aspects of sex role identity: physical appearance, personality, mannerisms, speech, interests, and habits. One's social sex role is to some degree independent of one's physical, gender, or sexual orientation identity. To illustrate this independence, Storms [34] used an androgeny scale (one that assesses masculinity and femininity independently) and found no significant differences in the masculinity and femininity of heterosexuals, homosexuals, and bisexuals.

Bem [3] and others have recognized masculinity and femininity are not simply opposite ends of a continuum. It is important to measure masculinity and femininity separately for each individual to indicate the strength or weakness of traditional masculine or feminine identification.

Finally, one's *sexual orientation identity* refers to the sexual or affectional interest in the same or opposite sex. DeCecco [15] says "Sexual orientation refers to the individual's physical activity with, interpersonal affection for, and erotic fantasies about members of the same or opposite biological sex." As stated previously, Kinsey *et al.* [21] suggested the use of a 7-point scale to describe one's sexual orientation. Several refinements to this classification system have been offered. First, sexual orientation is more complex than a singular scale with homosexual and heterosexual at its end points. Behavior, fantasy, and emotional or affectional attractions need to be rated separately because they are not always uniform in any individual. These ratings are not the same over different points in time, either. Finally, DeCecco [15] argues that sexual orientation can be best conceptualized when using two continuous scales, one for homosexual and one for heterosexual behavior. He thinks this eliminates the notion that the bisexual positions are simply a mixture of the two extreme components.

Sexual identity can now be viewed in its complexity. We need to be careful how the term bisexuality is being used in terms of sexual identity. Most commonly, it is understood as a description of the sexual orientation identity of the individual and should not be confused with gender, physical, or sex-role identity. The remainder of this paper shall limit itself to a discussion of bisexuality in terms of sexual orientation identity.

BISEXUALITY, DOES IT EXIST?

Even though we have a conceptual understanding of the existence of bisexuality, there are still many skeptics of its existence. McDonald [25] has identified four positions regarding bisexuality: (1) those who believe bisexuality is a real or natural state; (2) those who see it as transitory; (3) those who believe it to be transitional, and (4) those who see it as a denial of one's homosexuality. So, for most, bisexuality is viewed as a pseudo-orientation. However, let us go back to Kinsey *et al.* [21]:

> Since only 50% of the population (males) is exclusively heterosexual throughout its life, and since only four percent of the population is exclusively homosexual throughout its life, it appears that nearly half (46%) of the population engages in both heterosexual and homosexual activities, or reacts to persons of both sexes, in the course of their adult lives ([21], p. 656).

These may be the best figures available for the true incidence of bisexuality. In looking at research samples of homosexuals, it is obvious that many of these individuals could be classified as bisexuals. In their American sample, Weinberg and Williams [36] found that 49.4% reported some degree of bisexuality. Humphreys [20] in his study of men who frequented public restrooms for homosexual contact reported that 54% of his sample were currently married. Kinsey *et al.* [21] estimated that 2–10% of married men have had homosexual activity during their marriage.

In Bell and Weinberg's [2] extensive study of homosexuality, 3/4 of the males and between 2/3–3/5 of the women were exclusively homosexual in behavior. In both samples, the study participants were less apt to consider themselves exclusively homosexual in their 'sexual feelings'. McDonald [25] suggested that between 26–39% of the Bell and Weinberg sample reported their sexual behavior to be bisexual and that 38–55% reported their sexual feelings to be bisexual.

There is some indication from the Bell and Weinberg study and others that bisexuality is more common among females than males [1, 19, 37, 39]. McDonald [25] has also suggested that bisexual behavior may have recently increased for women. Blumstein and Schwartz [5] noted that the amount of bisexual behavior among women may have increased as a result of more women becoming active in the women's liberation movement. There is also speculation that women may find more cultural and ideological support for bisexuality than men. Women find support through feminist thought and the fact that bisexual behavior is a natural extension of societal acceptance of female affectionate behavior [29]. Adult males exploring homosexual relationships are not as likely to find this kind of support. Men fear an eroding of their masculinity, virility, and social acceptability. As a result, Paul [29] has found that women report initial homosexual experiences much less traumatic than men.

If based solely on the Kinsey *et al.* research, Klein [22] has estimated that 30–45% of the males and 15–35% of the females are definable as bisexual (Kinsey 1–5). This means that between 30–40 million people in the United States may be classified as bisexual – and, maybe more. We have just indicated that there may be many more women bisexuals than men, contrary to Kinsey *et al.* findings. So, the numbers, while not definitive, do speak for themselves. Bisexuality exists.

WHAT ARE THE PROBLEMS FACED BY BISEXUALS?

Certainly, bisexuals must first face the problem of the lack of public acceptance for their existence. There is no 'bi-movement', no society of their own, and support for a bisexual lifestyle or identity is definitely lacking.

The bisexual individual also must face the myths which Klein [22] has identified: the bisexual is by definition neurotic; the bisexual cannot love deeply; the bisexual is mixed up and can't make up his/her mind; and the bisexual is hypersexual and sex crazy.

This lack of societal acceptance and perpetuation of mythology comes from both 'gays and straights'. Both are uncomfortable with bisexual individuals. What is the threat? The threat appears to be similar to 'homophobia' for heterosexual people. The existence of an 'opposite' sexual orientation can be threatening to an individual with a fragile and insecure sexual identity. So, too, 'biphobia' exists in individuals who are easily threatened and the underpinnings of their sexual orientation is tenuous and shaky.

In her study of 150 bisexual men and women in Great Britain, Charlotte Wolf [39] indicated that her participants clearly saw bisexuality as a social disadvantage. This was counterbalanced by the fact that these individuals also saw bisexuality as an emotional, intellectual, and creative advantage.

Strong and positive identity development is still difficult under these circumstances. The pressure is to form a heterosexual or homosexual identity. Gay liberation has freed many homosexuals to form a positive sexual orientation identity, however, failure to assume a homosexual identity is now viewed as contrary to gay liberation [4].

One of the obvious problems faced by bisexual individuals is the issue of exclusivity or monogamy in relationships. With a desire to have both homosexual and heterosexual expression, one partner cannot, by nature, provide both outlets. Wayson points out "there is a clear price to be paid and that price may well be at the loss of a potentially stable, enduring relationship with a woman (or a man)" ([35], p. 108). However, lack of exclusivity in relationships should not be viewed as a necessary problem for bisexual individuals. As Klein [22] indicated, many individuals do not simultaneously pursue male and female relationships. This does not preclude a bisexual identity. These issues need to be treated separately. Heterosexual and homosexual individuals face the same dilemmas. No one person satisfies all erotic or emotional interests. Every individual, then, faces the issue of pursuing other relationships within the context of a primary relationship or remaining exclusive and sublimating those outside interests. The difficulty of this sublimation may be more difficult for the bisexual, however, because of the physical differences between males and females, but this has not been proven.

Sexual expression may have some correlation with positive sexual orientation identity but it is by no means a one-to-one relationship. Individuals with no homosexual expression still are able to identify strongly as a bisexual and feel positive about their sexual orientation identity.

The other problems faced by bisexuals have to do with marriage and children. Homosexual expression within heterosexual marriage has meant inherent difficulties (cf. [9, 11, 12]). Striving for bisexual expression leads to inevitable difficulties although certainly not inevitable dissolution of the marriage. Some models of integration have been developed for this situation. Interestingly enough, this seems to be more viable for bisexual men than women [12].

A related problem is whether bisexual individuals can be good parents. There are popular fears that homosexuality is contagious, that the children are in danger of being molested by the bisexual parent(s), or that the bisexual parent(s) create(s) an inherently unhealthy environment, and that the children will suffer ridicule from other children and the community. There has been no evidence that these myths are true on a global level and are only true in specific instances. The same can be said about heterosexual parents. Some, but certainly not the majority, impose their sexual attitudes on their children, molest their children, create unhealthy environments, and inflict ridicule on their children because of their socially-unacceptable behavior (just as some heterosexual parents do).

It may be that bisexual individuals may be less satisfied with their sexual orientation identity because of the aforementioned problems. Certainly, societal intolerance must play a role in the lack of self acceptance and satisfaction. Because of this social intolerance, bisexual individuals have to make difficult choices: to remain single, marry someone who accepts your bisexuality, take the risk of marriage with deception, marry and abstain from homosexual involvements, or live in a homosexual relationship under some of these same conditions [25]. In order to survive these problems, the bisexual individual needs to have a high tolerance for ambiguity and an internal locus of control. Paul [29] concludes that ". . . the bisexual needs to adopt a broader, better integrated perspective in human sexuality and social relationships."

PSYCHOLOGICAL ADJUSTMENT

Facing these problems is certainly not easy. What effect does this have on psychological adjustment? As indicated, the bisexual individual is vulnerable to the confusion and stress arising from a dual sense of self and the attempts to merge both worlds with a single identity [29].

However, researchers such as Weinberg and Williams [36] have not found bisexual men reporting greater psychological difficulties than heterosexual or homosexual men. While no overall differences were found, bisexual individuals were more likely to report feelings of guilt, shame, or anxiety over being homosexual. But these authors did not find these problems to generalize to other psychological problems. The authors cited this as the evidence for the human capacity for adaptation and compartmentalization.

In another and more recent study of bisexuality, Wayson [35] found no significant differences between a sample of heterosexual, bisexual, and homosexual men on a measure of psychological adjustment. In fact, while not statistically significant, the bisexuals had somewhat higher scores of self esteem than the other samples.

In Bode's [6] study of female bisexuality, she found that the majority of the women were well educated, single, and came from unbroken nuclear families with parents who were behaviorally heterosexual. Most of the women were comfortable with their sexual lifestyle. Their attitude was that if the public was disconcerted about bisexuality then the national attitude should change.

In another study of female bisexuality, La Torre and Wendenberg [23] found the bisexual women in her study displayed the same psychological characteristics as did a sample of heterosexual and homosexual women. Overall, not much difference could be found between the samples of heterosexual, homosexual and bisexual individuals.

However, differences can be found within samples, which indicates other variables are probably more powerful in affecting psychological adjustment than simply an individual's sexual orientation. For example, Rand, Graham, and Rawlings [31] found that the psychological health of lesbian mothers correlated positively with disclosure of their sexual orientation to employers, ex-husbands, children and with the amount of feminist orientation. Also, among bisexual married men, psychological health and marital adjustment and satisfaction seems to be positively correlated with high levels of income, education, open-communication, maintenance of satisfactory sexual relations, participation in psychotherapy, and lower amounts of homosexual activity (cf. [7, 12, 38]). As with homosexuality, this identity does not inevitably lead to psychological maladjustment. Certain factors lead to better psychological functioning. The same is true for the bisexual. Different lifestyles or ways of bisexual expression may lead to more or less well-adjusted psychological functioning.

IS THERE A TYPICAL BISEXUAL LIFESTYLE?

There is no more a descriptive bisexual lifestyle than there is a heterosexual or homosexual lifestyle. As a result of this fact, researchers such as Bell and Weinberg [2] refer to homosexuality as *homosexualities* and recognize a number of different lifestyles, patterns, and adjustments. This is certainly also true for bisexual individuals. As stated, we have already a number of different typologies for bisexual individuals (e.g. [22]). No single lifestyle can be found [4, 5].

Are bisexual individuals more sexually active? This is unresearched at this time.

IS THERE A PATTERN OF BISEXUAL IDENTITY DEVELOPMENT?

Stages of identity development for homosexual individuals have been offered by a number of researchers (e.g. [8, 10, 14, 30]). I have suggested a series of five stages for individuals with a predominant same-sex orientation. It is not clear whether bisexuals follow a similar developmental path or not. In the first stage, *pre-coming out*, same-sex feelings are unconscious. In the next stage, *coming out*, the individual consciously acknowledges that same-sex feelings exist. This acknowledgement is through self awareness as well as acknowledgement to others. The developmental task of this stage is to begin to develop some positive feelings about self, contrary to the negative feelings felt during the *pre-coming out* stage. *Exploration* is the stage of adolescence. The individual is learning new interpersonal and intimacy skills, further defining the self, and defining self as attractive. Oftentimes, this stage occurs outside of chronological adolescence and therefore is confused with psychopathology. *First relationships* is the stage when the individual works on establishing and maintaining intimate, romantic relationships. First attempts are usually unsuccessful. Many relationship skills are learned and developed in this stage. Finally, a fifth stage of *integration* occurs when many of the developmental tasks of earlier stages are accomplished. These stages are further defined and illustrated in another paper [10].

Obviously, many of these developmental stages are not the exclusive domain of individuals who are predominantly or exclusively homosexual. Any amount of same-sex feelings may lead to this or a similar developmental pattern.

This model recognizes that every individual does not follow each stage and evolve through all of them. Some never experience identity integration. The model also recognizes that the developmental history of most individuals is probably more chaotic, fluid, and complex than the model describes, but the framework is still useful if used in a flexible manner. While the model has been substantiated to some degree through research on individuals with predominantly same-sex orientations, it has not been substantiated on a sample of bisexual individuals. We have very little data on identity development of bisexuals. In Klein's study [22] of 144 individuals (103 males and 41 females) attending the Bisexual Forum in New York City, the average age of the males was 28.5 and 32.0 for the females. The average age the participants identified themselves as bisexual was 24.1 (males) and 24.4 (females). While the bisexual males experienced first opposite-sex experiences about the same time as the females (females – 16.0; males – 17.8), the females experienced their first same-sex experience much later than males (females – 23.0; males – 15.5). Males, therefore, developed their bisexual identity approximately 9 years after their first homosexual experiences and the females developed their identity after only one year. This indicates there may be a significant difference in the dynamics and course of bisexual identity formation for females than males. Women appear to progress in terms of identity development more quickly even though their first homosexual experience occurs later [29]. However, Blumstein and Schwartz [4, 5], in their study of bisexual men and women, found that males were more likely to maintain an uninterrupted bisexual self definition once the label was adopted than were the female bisexuals. These authors suggested that female sexual identity was more tied to the sex of the current partner and that bisexuality as a label for females was a difficult concept because of women's greater commitment to the ideal of sexual monogamy.

The differences between men and women need to be explored further. However, basic similarities, I am sure, remain. For example, the importance of the *coming out* stage is rather universal. Bozett [7] described how gay (bisexual?) fathers integrated their dual identities by gradual disclosure to homosexual and heterosexual intimates. These men achieved greater integration identity by *integrative sanctioning*, "a process allowing others to confirm, reinforce, or indicate acceptance of the dual identities." The same may be true for men and women with bisexual orientations.

The process may be inhibited, as I have mentioned, because of the lack of societal support and/or a lack of a bisexual liberation movement. Another fact remains: there is still little sign of a bisexual movement emerging. Margaret Mead [26] has said it seems unlikely a bisexual liberation movement will emerge because of the fact that bisexual men and women do not form a distinct group and that society still does not recognize bisexuality as a form of behavior, normal, or abnormal. This was true, however, for homosexuality before the riots at Stonewall in Greenwich Village in 1969. Time will tell.

Without this societal support, the identity development and achievement of identity integration is a difficult task for most. I fear many suffer or simply adopt a heterosexual or homosexual identity because more societal sanctioning exists for these identities.

HOW DO I KNOW WHO IS BISEXUAL AND WHO IS HETEROSEXUAL OR HOMOSEXUAL?

The problems of definition have been reviewed extensively. Recognizing these difficulties, practical tools for assessment of sexual orientation will be offered. Minimally, one can use the unidimensional Kinsey Scale (0–6) with heterosexuality and homosexuality at its end points. As has been criticized, the one scale is probably insufficient. At least several Kinsey scales should be used to measure behavior, erotic fantasies, and emotional attachments. The clinical tool which I have found easiest and more sophisticated gives a well-balanced picture of the patients' sexual orientation. Instead of individuals rating themselves on a line, circle graphs are used, which individuals can divide into percentages – or slices of a pie. Three basic dimensions are used: sexual behavior, erotic fantasies, and emotional attachments. A fourth open circle can be used for the patient to use for any other dimension that seems relevant to him or her (see Figure 2).

Normally, the patient is asked to fill out one set of circle graphs for the past and then to fill out a set describing what they want in the future or their goals of counseling. If different time periods seem to be relevant given his or her sexual history, other sets of circle graphs can be filled out. It is not necessary to be rigid in using these graphs because their only utility is clinical rather than for research purposes.

CASE EXAMPLES

In Case 1 (Fig. 2a) this 23-year-old male patient had most of his sexual experiences with men and had exclusive same-sex fantasies. He had emotional attachments with men and women. He wanted to eliminate all behavior and fantasies of men and to be sexually active with women. This is a good illustration of ego-dystonic homosexuality. Further counseling was recommended to resolve this conflict in sexual orientation.

Case 2 (Fig. 2b) illustrates a 17-year-old woman who had a one-time sexual experience with another young woman. This experience frightened her, and she was concerned she might be lesbian. By illustrating her overall pattern of sexual behavior, fantasy and emotional attachments, she was relieved of her homosexual panic. She was also reassured with information concerning the universality of bisexuality in

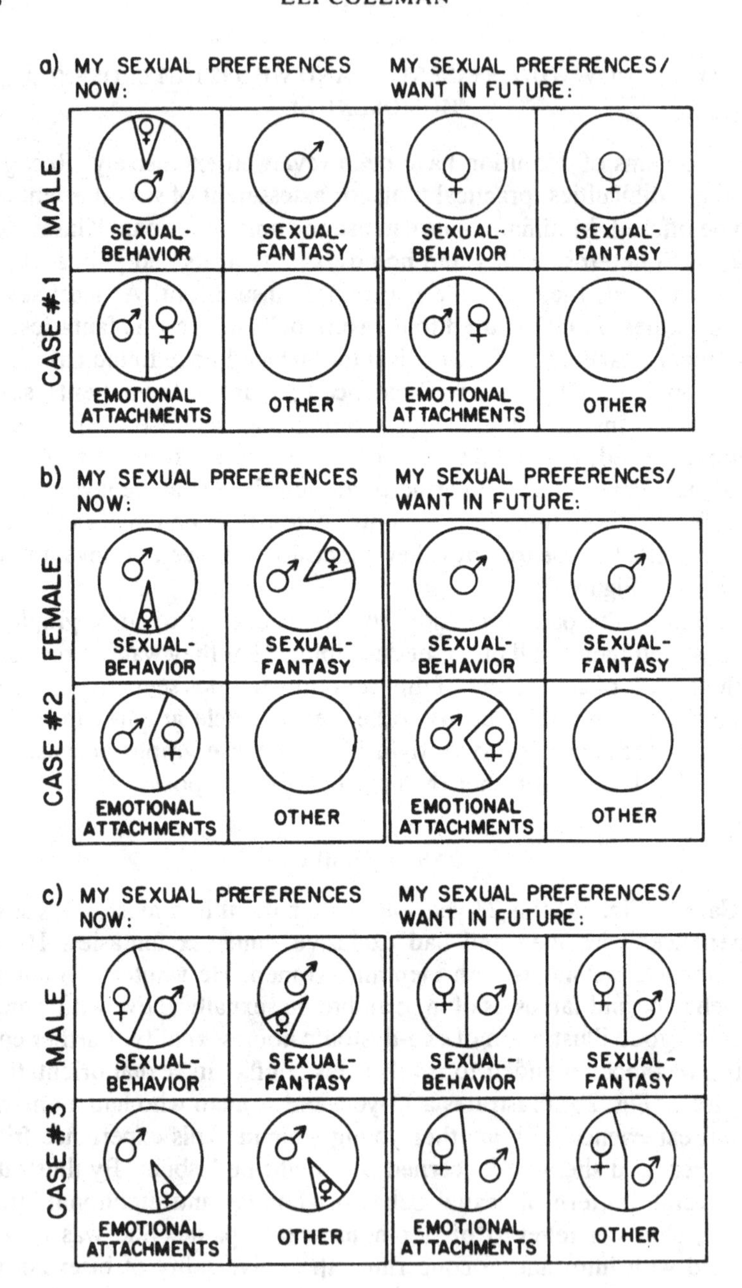
a) MY SEXUAL PREFERENCES NOW:
MY SEXUAL PREFERENCES/ WANT IN FUTURE:
CASE #1 MALE
SEXUAL-BEHAVIOR
SEXUAL-FANTASY
SEXUAL-BEHAVIOR
SEXUAL-FANTASY
EMOTIONAL ATTACHMENTS
OTHER
EMOTIONAL ATTACHMENTS
OTHER
b) MY SEXUAL PREFERENCES NOW:
MY SEXUAL PREFERENCES/ WANT IN FUTURE:
CASE #2 FEMALE
SEXUAL-BEHAVIOR
SEXUAL-FANTASY
SEXUAL-BEHAVIOR
SEXUAL-FANTASY
EMOTIONAL ATTACHMENTS
OTHER
EMOTIONAL ATTACHMENTS
OTHER
c) MY SEXUAL PREFERENCES NOW:
MY SEXUAL PREFERENCES/ WANT IN FUTURE:
CASE #3 MALE
SEXUAL-BEHAVIOR
SEXUAL-FANTASY
SEXUAL-BEHAVIOR
SEXUAL-FANTASY
EMOTIONAL ATTACHMENTS
OTHER
EMOTIONAL ATTACHMENTS
OTHER

Fig. 2.

many of us and that this does not inevitably signal future exclusive homosexuality.

In the third case (Fig. 2c) illustrating a 20-year-old male, the circle graphs indicate that he has had almost equal same-sex and opposite-sex sexual experiences. His fantasies are more exclusively male oriented and he also becomes more emotionally involved with males. When asked about his future, he expressed an interest in maintaining a bisexual lifestyle. He was quite content with his orientation but simply needed reassurance that his same-sex feelings were normal. We discussed the difficulty of maintaining a bisexual lifestyle and the many options available. He did not see any conflict with his sexual orientation and felt he could handle the future difficulties. No further counseling was recommended. In this case, the patient used the fourth circle to describe his feelings of security while engaging in sexual activity with men and women.

These three cases illustrate the usefulness of this assessment device. Both the physician and the patient can get a better picture of the patient's sexual orientation. The filling out of these circle graphs is non-threatening to the physician and patient and serves as a stimulus for further discussion and clarification.

Filling out the second set of circle graphs to describe future objectives indicates conflicts or satisfactions in sexual orientation. The more discrepancies between past and future usually indicates some conflict and the need for further counseling and therapy. As illustrated in the case examples, many sexual orientation concerns can be handled easily with reassurance and information in a short amount of time. When conflicts are evident, further counseling by the physician is indicated, or referral to a qualified psychotherapist.

More elaborate assessment devices have been offered by Fred Klein as described by Wayson [35]. Individuals are asked to define who they are attracted to, sexual fantasies, sexual behavior, emotional dependence, socialization pattern, lifestyle, and self identification. These are rated on a seven-point scale for past, present, and ideal situations.

Since a particular rating on any of the scales is not definitive of sexual orientation, self assessment may still be the best indicator. On patient, social-history intake forms, I recommend that patients be asked to identify their sexual orientation: exclusively heterosexual, predominantly heterosexual, bisexual, predominantly homosexual, exclusively homosexual; unsure. The professional, however, must take into account the pressures on the individual not to develop a bisexual identity. And, for others, they might find a bisexual identity a convenient escape from adopting a homosexual identity which may more closely reflect the person's true sexual orientation. Therapists need to help their patients define themselves and recognize the value and complexity of their sexual orientation.

Therefore, my approach has been to help individuals assess, understand, accept and value their sexual orientation whatever that is. Beyond acceptance and valuing, the individual faces the task of being who he or she is in a predominantly heterosexual society. Survival skills need to be learned. It is a challenge for anyone who does not fit into a discrete category to feel good about him/herself and to develop a positive, mature and integrated identity. There are few models of survival outside of heterosexual marriage and life in the gay community. Models based upon diversity and complexity of sexual orientation need to be developed.

CONCLUSION

I hope that a further appreciation of the complexity of sexual orientation has been reached. The labels we have – homosexual, bisexual, heterosexual seem meaningless when we realize the complexity. The words homosexual and heterosexual seem the most limiting. That is why I often use the words predominantly homosexual or predominantly heterosexual. I will use the word bisexual probably more freely – not to describe someone in the middle of every conceivable Kinsey continuum but someone who has a mixture of same-sex and opposite feelings, attractions, and experiences. And, I do believe that many fall into this category. One of the ways our culture will get over their homo- or bi-phobia is if we recognize the continuum concept – that we are not all straight or gay and that we do not have two opposing camps. We can eliminate the 'homosexual panic' which many predominantly heterosexual individuals suffer which is one of the main sources of the homo- or bi-phobia.

Conversely, I work with many gay-identified individuals who are experiencing 'heterosexual panic'. Some have awakened some morning after having dreamed they just made love to a woman and fear they have made a mistake in terms of their lifestyle. "Maybe I'm really straight and my mother was right." This rigidity in thinking is perpetuated by the scientific community and is unnecessary given our scientific knowledge.

We also need to explore more the relationships among emotions, fantasies, and the physical aspects of sex. McDonald [25] concludes we need to explore how different societies and cultures determine which aspects of sexuality are to be suppressed, suffered, and enjoyed. There is still much to be learned. In the meantime, the information we do have needs to be understood by all and more recognition be given for the simple reality that bisexuality exists and is a normal part of human sexual behavior.

University of Minnesota,
Minneapolis, Minnesota,
U.S.A.

BIBLIOGRAPHY

1. Altman, D.: 1983, *The Homosexualization of America*, Beacon Press, Boston.
2. Bell, A. P. and Weinberg, M. S.: 1978, *Homosexualities: A Study of Diversity Among Men and Women*, Simon and Schuster, New York.
3. Bem, S.: 1974, 'The Psychological Measurement of Androgeny', *Journal of Consulting and Clinical Psychology* **42**, 155–162.
4. Blumstein, P. W. and Schwartz, P.: 1976, 'Bisexuality in Men', *Urban Life* **5**(3), 339–358.
5. Blumstein, P. W. and Schwartz, P.: 1976, 'Bisexuality in Women', *Archives of Sexual Behavior* **5**, 171–181.
6. Bode, J.: 1976, *View from Another Closet: Exploring Bisexuality in Women*, Hawthorne, New York.
7. Bozett, F.: 1981, 'Gay Fathers: Identity Conflict Resolution Through Integrative Sanctioning', *Alternative Lifestyles* **4**(1), 90–101.
8. Cass, V. A.: 1970, 'Homosexual Identity Formation: A Theoretical Model', *Journal of Homosexuality* **4**, 219–235.
9. Coleman, E.: 1982, 'Bisexual and Gay Men in Heterosexual Marriage', *Journal of Homosexuality* **7**, 93–103.
10. Coleman, E.: 1982, 'Developmental Stages of the Coming Out Process', *Journal of Homosexuality* **7**, 31–43.
11. Coleman, E.: 1985, 'Bisexual Women and Lesbians in Heterosexual marriage', *Journal of Homosexuality* **11**(1/2), 87–113.
12. Coleman, E.: 1985, 'Integration of Male Bisexuality and Marriage', *Journal of Homosexuality* **11**(1/2), 189–207.
13. Cook, K.: 1983, 'The Playboy Readers Sex Survey: Part Three', *Playboy* (May), 126–126; 136; 210–220.
14. Dank, B. M.: 1971, 'Coming Out in the Gay World', *Psychiatry* **34**, 180–197.
15. DeCecco, J.: 1981, 'Definition and Meaning of Sexual Orientation', *Journal of Homosexuality* **6**(4), 51–67.

16. Ford, C. S. and Beach, F. A.: 1951, *Patterns of Sexual Behavior*, Harper and Row, New York.
17. Freud, S.: 1962, *Three Essays on the Theory of Sexuality*, Basic Books, New York. (Originally published in 1905).
18. Gagnon, J. H.: 1977, *Human Sexualities*, Scott, Foresman, and Co., Glenview.
19. Hite, S.: 1976, *The Hite Report*, Macmillan, New York.
20. Humphreys, L.: 1975, *Tearoom Trade: Impersonal Sex in Public Places*, Aldine, Chicago.
21. Kinsey, A., Pomeroy, W., and Martin, C. W.: 1948, *Sexual Behavior in the Human Male*, W. B. Saunders, Philadelphia.
22. Klein, F.: 1978, *The Bisexual Option*, Arbor House, New York.
23. LaTorre, R. A., and Wendenberg, K.: 1983, 'Psychological Characteristics of Bisexual, Heterosexual and Homosexual Women', *Journal of Homosexuality* **9**(1), 87–97.
24. Masters, W. H. and Johnson, V. E.: 1979, *Homosexuality in Perspective*, Little, Brown and Company, Boston.
25. McDonald, A. P.: 1981, 'Bisexuality: Some Comments on Research and Theory', *Journal of Homosexuality* **6**(3), 21–35.
26. Mead, M.: 1975, 'Bisexuality: What's It All About?', *Redbook* (January), 29–31.
27. Money, L.: 1984, 'Bisexuality and Homosexuality', *Sexual Medicine Today* (February), 24.
28. Money, J. and Ehrhardt, A. A.: 1972, *Man and Woman, Boy and Girl: The Differentiation and Dimorphism of Gender Identity from Conception to Maturity*, Johns Hopkins University Press, Baltimore.
29. Paul, J. P.: 1983–84, 'The Bisexual Identity: An Idea Without Social Recognition', *Journal of Homosexuality* **9**(2/3), 45–63.
30. Plummer, K.: 1975, *Sexual Stigma: An Interactionist Account*, Routledge and Kegan Paul, London.
31. Rand, C., Graham, D. L. R., and Rawlings, E. I.: 1982, 'Psychological Health and Factors the Court Seeks to Control in Mother Custody Trials', *Journal of Homosexuality* **8**(1), 27–39.
32. Shively, M. and DeCecco, J. P.: 1977, 'Components of Sexual Identity', *Journal of Homosexuality* **3**(1), 41–48.
33. Shively, M., Rudolph, J., and DeCecco, J. P.: 1978, 'The Identification of the Social Sex-role Stereotypes', *Journal of Homosexuality* **3**(3), 225–233.
34. Storms, M. D.: 1980, 'Theories of Sexual Orientation', *Journal of Personality and Social Psychology* **38**, 783–792.
35. Wayson, P. D.: 1983, 'A Study of Personality Variables in Males as They Relate to Differences in Sexual Orientation', Doctoral Dissertation, California School of Professional Psychology, San Diego.
36. Weinberg, M. S. and Williams, C. J.: 1974, *Male Homosexuals: Their Problems and Adaptations*, Oxford University Press, New York.
37. West, D. J.: 1977, *Homosexuality Re-examined*, University of Minnesota Press, Minneapolis.
38. Wolf, T. J.: 1982, 'Selected Psychological and Sociological Aspects of Male Homosexual Behavior in Marriage', Doctoral Dissertation, United States International University, San Diego.
39. Wolff, C.: 1977, *Bisexuality*, Quartet, London.

PETER ROBERT BREGGIN

SEX AND LOVE: SEXUAL DYSFUNCTION AS A SPIRITUAL DISORDER

SEX IN ITS WHOLE CONTEXT

The attempt to focus on 'sexual dysfunction' outside its psychological, social, and philosophical context is like trying to define a southern black man's difficulties getting to the polling booth as a 'walking dysfunction' or perhaps like talking about a priest's struggle with sexual continence as an 'emission control' problem.

As much or more than any other 'biological' function, sexuality for each and every human being is tied to the entire history of human sexual relations, which means, to the entire history of *human relations*. When a woman and man get into bed together, Freud claimed, there are always two other people present – the parents of the opposite sex. More comprehensively, when a woman and man get into bed together they bring with them the collective impact of the billions of other souls whose experiences throughout time have helped to create our attitudes toward sex and love.

FRIGIDITY IN SEARCH OF A CURE

Consider the situation of a forty-five-year-old white married Catholic woman raised in northeastern America in a traditional family who turns to the yellow pages to find a therapist to help her overcome her sexual 'frigidity.' Nothing special here: frigidity is said to be the most frequent 'sexual complaint' among women. If we believe that sex therapy is a behavioral or medical science in any sense, then the therapist should be able to offer a somewhat standardized therapy for the problem.

But even in regard to this most frequent sexual complaint, frigidity, the entire therapeutic enterprise, and hence the future prospects and perspectives of the client, will be arbitrarily determined by the particular attitudes toward sex embodied in the therapist. No matter how client-oriented he may be, his questions and his acknowledgements will be guided by his personal and/or professional attitudes and philosophies pertaining to women, to middle-aged women in northeastern America,

E. E. Shelp (ed.), Sexuality and Medicine, Vol. I, 243–266.

to people from traditional families, and of course, to Catholics and Catholicism. And to the extent that he tries to impose a 'value free' or 'objective' viewpoint, he only succeeds in imposing a scientistic (not scientific) or simple-minded behavioral philosophy.

As I have described in my books and articles [6–9, 11, 12], all therapy is essentially applied ethics and politics. Therapy may be called "applied utopian ethics and politics," [8, 9] since it reflects the therapist's attempt to create an ideal problem solving environment, and since this ideal must be based on overt or covert principles about human nature and human relations.

Insight Therapy with an Anti-Catholic Bias: Let us return to our forty-five-year-old Catholic client with a complaint of frigidity as she seeks help through the yellow pages under psychiatry. If she by chance selects a psychiatrist with a 'liberal' attitude toward sex, he has very likely become convinced that Catholicism, Catholic school, and the typical Catholic family promote medieval attitudes toward sexuality that typically result in sexual suppression. A therapeutic inquiry from this orientation would very possibly lead to a more 'liberated' sexual attitude on the client's part, and perhaps to disillusionment with the church and to divorce, as well as to other vast changes in her outlook and life.

Traditional Catholic Therapy: Now suppose that our client picks from the yellow pages a psychiatrist who turns out to be a dogmatic old-school Catholic, from whom she learns directly or indirectly that her sexual expectations are too high, that the Pope himself has condemned lust, and that she should redouble her religious devotions as well as her devotion to her husband. The outcome of this experience would differ vastly from her more 'liberal' therapy.

Insight Therapy with a Spiritual Bias: As a third alternative, imagine that our middle-aged Catholic client chances upon a therapist with a more spiritual and loving orientation toward sex, a viewpoint shared by many humanists and by some Catholics as well (e.g., the charismatic priest, Richard Rohr [38]). She discovers that she has not merely suppressed her sexuality, but more importantly that she has suppressed the God-given connection between sex, romantic love, shared Christian values, and a personal awareness of God's redeeming love for all his human children. Perhaps she joins a charismatic Catholic group that inspires her with the love of Jesus, through which she ultimately includes her husband in her revitalized romantic passion.

Feminist Therapy: Now suppose that our client picks a woman psychiatrist from the yellow pages who turns out to be a feminist and who advises her to read Kate Millet's *Sexual Politics* [31] and Betty Friedan's *The Feminine Mystique* [21]. The books prove to be a revelation of a different sort to this Catholic woman, who begins a re-evaluation of her experience in the patriarchal Church as well as in the patriarchal families in which she grew up and into which she ultimately married. Her 'frigidity,' she discovers, was a covert rebellion against the traditional morality that commanded her to submit to her insensitive husband. The implications for her sexual life, and indeed for her entire life, are vast.

Psychoanalytic Therapy: Or suppose this middle-age Catholic woman happens by chance to find herself in the office of a classically trained psychoanalyst who 'objectively' informs her that she needs analysis four to five times a week for several years in order to deal with her problems. I leave to the reader's imagination the infinite potential outcomes of this particular approach, with the various colorations achieved depending on the personality and the particular therapeutic style of this man with whom she will now spend so many hundreds and even thousands of hours.

Behavioral Sex Therapy: Now consider the possibility that the name in the yellow pages happens to be that of a behaviorally oriented sex therapist who does not investigate her past in any depth, who has no concept of a spiritual connection between sex, romantic love, and Christianity, and who has never read a feminist book. If this is a typical sex therapist, the issue of 'love' may never be raised or addressed in any manner whatsoever, as in the instance of a client who came to me complaining of 'incurable frigidity' after having spent two disappointing weeks at a well-known sex therapy clinic. After only a few minutes in my office, I asked her if she loved her husband. She replied in astonishment, "Do you think that has something to do with it?" – after which she spontaneously declared, "No, I've never loved him. My goodness, do you think that could be it?"

Medically Oriented Sex Therapy: There is some chance that our client might end picking a biologically oriented sex therapist who is an expert on hormonal deficiencies. He takes a medical history and discovers that this lady's malady began almost simultaneously with the prodromal signs of her menopause. He immediately orders a blood test to check for hormonal levels. He receives the lab results and, with exemplary bedside manner, explains to the client the nature of her hormonal imbalance. Sure enough she soon reports that there has been some increased desire for sex on her part as a result of hormonal replacement therapy.

Her husband in particular thinks the doctor is a genius.

Ordinary Techno-Psychiatry: I have left to the last one of the most potentially dreadful alternatives for this middle-aged lady who complains of frigidity. She might select an ordinary psychiatrist from the yellow pages – one who believes what he has been taught about clinical diagnoses, chemical imbalances, and drug therapy. This sophisticated psychiatrist might see beneath her presenting complaint and diagnose her 'depressed affect.' It does not occur to him that her depression is the inevitable product of her personal problems, but instead concludes her sexual and family problems are caused by her depression – and he gives her an antidepressant for her alleged endogenous depression. After taking the antidepressant for two weeks, she notices a loss of energy, memory problems, occasional confusion, and difficulties concentrating (for an analysis of brain-disabling drug effects, see [14]). These drug-induced symptoms are mistaken for a worsening of her depression, and she is hospitalized for electroshock treatment. She never recovers her mental faculties as the shock makes permanent the dysfunction originally caused by the antidepressant (for an analysis of electroshock effects, see [10]). She returns home to become a drone at housework who no longer can find solace in her religion or her family. Later on a neurologist performs a sophisticated computerized X-ray of her brain called the CAT scan, correctly diagnosing her as suffering from shrinkage of her brain. Her psychiatrist considers this proof that she has a 'mental illness' of biological origin. No one realizes that her irreversible dementia is caused by drugs and electroshock.

THE CLIENT'S VULNERABILITY

The client's response may not be too helpful in deciding what approach is best. Often the client seeks help precisely because he/she is 'at the end of the rope,' and is desperate for any kind of help. In my own experience as a psychotherapist, and also as a physician, the patient or client seeks help at exactly that moment when he/she feels most personally helpless [13], hoping for a simple solution and for decisive direction from a trustworthy authority. She is vulnerable and suggestible. Within limits that vary from person to person, she is likely to accept a wide variety of approaches to her problem. The outcomes may vary greatly depending on the therapy; but she may find something to feel grateful for in an infinite number of conflicting alternatives. And if she has been damaged by the treatment, her therapist is likely to blame the new symptoms on her 'mental illness' [13].

SEX: THE VIEW FROM ABOVE (AND BELOW)

My own experience with myself, my friends, and my clients has led me to make a direct connection between consistently good sexual experiences and romantic love. In the following sections I will present a number of differing viewpoints on the relationship between sex and love in the hope of stimulating a greater appreciation of the subject. Whether you come out 'for or against' romantic love, you will better understand the feelings of most of your clients if you grasp the importance of our conflicting attitudes toward romantic love.

The Origins of Romantic Love

The notion is frequently expressed in the literature that romantic love is a relatively modern concept. Claiming that romantic love could not have developed without modern capitalism and individualism, Branden [5] argues that romantic love was seen as an illness by the Greeks and Romans, and utterly denigrated by later Christian morality. An advocate of romantic love, he tells us:

> Christianity upheld to men and women an ideal of love that was consistently selfless and nonsexual. Love and sex were, in effect, proclaimed to stand at opposite poles: the source of love was God; the source of sex was, in effect, the devil ([5], p. 26).

But it can also be said that romantic love was nurtured by Christian theology, specifically by the conviction that each and every soul is imbued with God and therefore worthy of admiration and even worship. Thus Auden condemns romantic love as a sick view of life, attributable to the Christian idea of the precious worth of each and every soul. The early romantic myths of Don Juan and Tristan and Isolde are seen as "diseases of the Christian imagination":

> Whenever a married couple divorce because, having ceased to be a divine image to each other, they cannot endure the thought of having to love a real person no better than themselves, they are acting under the spell of the Tristan myth ([3], p. 29).

My own analysis finds some truth in both their views: many aspects of early Christianity were wholly antipathetic to erotic, romantic love; but the Christian idea of a divinely precious soul helped to foster romantic love, in preparation for its more full flowering in the atmosphere of modern individualism. I *disagree* with both of them about dating the earliest expressions of romantic love to relatively modern times. There are strong portrayals of romantic love in both the Hebrew Bible and classic mythology.

Romantic Love Among the Ancient Hebrews

Samson falls madly in love twice, each episode ending in disaster for himself and the Jewish community. As in many stories of passionate love, Samson's attachments break down traditional boundaries, in this case between the Jews and the Gentiles.

King David falls in love with Bathsheba at first sight, and though it is no exaggeration to say that there are 'problems in the relationship,' they do marry, and Bathsheba becomes an important advisor to the king.

The tale of Jacob and Rachel in Genesis is a classic story of romantic love. Jacob, the epitome of the independent, courageous man, strikes out on his own into the wilderness. He spies a beautiful woman at a well and instantly falls in love with her. He contracts to work for her father for seven years to gain her hand in marriage, and when he is tricked into marrying her sister instead, he contracts again to work for another seven years in order to marry her.

After Jacob and Rachel are finally married, his love continues to grow through the vicissitudes of life, other women, and children. Indeed, his first wife, Leah, is tormented because Jacob continues to love Rachel the most, even though Leah bears him children. This story has many elements of later chivalric love, including love at first sight, prolonged abstinence, and great labors to prove the love; and it ends happily in marriage – a much more optimistically modern conclusion than found in Tristan and Isolde or Romeo and Juliet, which end tragically.

Romantic Love in Classical Mythology

While the Greeks often described romantic love as madness, and the Romans treated it little better, we find the prototype of all future romantic love stories in the classical myth of Cupid and Psyche [16]. Although the story appeared fullblown in writing for the first time in

Latin in the second century, it is deeply rooted in more ancient mythology.

Psyche is such a beautiful young virgin that she draws the envy of Venus, who sends her son Cupid to infuse Psyche with compulsive love of some lowly being. Instead Cupid, the god, is smitten with this mortal woman, becomes discombobulated, sprinkles the wrong potion on her, stabs himself with his own arrow, and causes them to fall in love with each other for eternity.

Cupid flees. After years of loveless loneliness, Psyche is led to rediscover Cupid, who comes to her unseen at night to make sweet love to her. He is afraid to have her see that he is a god, and when she begs to look at him, he demures:

> 'Why should you wish to behold me?' he said; 'have you any doubt of my love? have you any wish ungratified? If you saw me, perhaps you would fear me, perhaps adore me, but all I ask of you is to love me. I would rather you would love me as an equal than adore me as a god' ([16], p. 71).

In this astonishing statement we find some of the most essential elements of modern romantic love – especially the desire to be loved as an equal, a human, rather than as a god.

When Cupid discovers that Psyche distrusts him, he declares he will abandon her forever. His parting words are "Love cannot dwell with suspicion." Psyche wanders the earth in despair seeking her lover, and her laborious tasks and adventures rival those of Tristan and pale those of Romeo. Cupid himself comes to her rescue and helps her through her last ordeal. Unlike a later bumbling Romeo, Cupid is not fooled when he finds his beloved stricken by a sleeping potion, and he revives her. Finally the great god Jupiter make Psyche immortal; Cupid and Psyche are bound together for eternity; and they consummate their love with a child named Pleasure.

This ancient tale has many elements that are usually attributed to modern romanticism, including the defiance of authority (the mother, Venus) as the two lovers overcome all obstacles in order to fulfill their love. It is also modern in the direct fulfillment of their sexual desires early in the story, followed by marriage and children. It is even feminist in its insistence on equality between the lovers, with the ultimate elevation of Psyche to equal status with Cupid as a divinity. The themes of faithfulness, monogamy, and trust also have a distinctly modern ring.

The myth beautifully dramatizes spiritual elements that are usually attributed to the Christian influence on romantic love. Psyche in Greek means soul. Psyche's beauty and her love for Cupid bestow divine qualities on her, and eventually she is literally made divine, an eternal soul, as a result of her love.

There are many other parallels between the myth of Cupid and Psyche, and later romantic love stories. They are primarily of interest to the literary historian. My point is that from ancient times sexual passion and love were closely connected in the great mythologies of Greece and the Hebrew Bible, and that these tales have come to influence the hopes and ideals of millions of persons throughout the ages. Today these hopes and ideals continue to impact on how people view themselves, their sexuality, and their partners in sex and love.

Freud's Humbug View of Romantic Love

Having established these ancient roots for the ideal, let us now turn to one of the great cynics of love, Sigmund Freud. Freud hated and feared sex. As Rieff [37], Fromm [24, 25], and others have documented, Freud used the alleged connection between sexual repression and civilized values as a method for degrading civilized values in general. Sexual repression and civilized activity were so closely connected in his mind that he declared that only the most powerful civilized individuals could continue to derive even the most meager satisfaction from sex.

Freud was especially derisive toward romantic love. It is not that Freud thought romantic love had nothing to offer; to the contrary he saw it as offering more pleasure and happiness than any other experience on earth. In *Civilization and Its Discontents* [22] he tells us "one of the forms in which love manifests itself, sexual love, gives us our most intense experience of an overwhelming pleasurable sensation and so furnishes a prototype for our strivings after happiness." So what is the problem? The pain of losing the loved one becomes too unbearable:

> he [the lover] becomes to a very dangerous degree dependent on a part of the outer world, namely, on his chosen love-object, and this exposes him to most painful sufferings if he is rejected by it or loses it through death or defection. The wise men of all ages have consequently warned us emphatically against this way of life; but in spite of all it retains its attraction for a great number of people ([22], p. 69).

Freud is so disillusioned about sex that he describes it as "being a function in process of becoming atrophied, just as organs like our teeth and our hair seem to be." For Freud, and I am tempted to say 'poor

Freud,' there is "something in the nature of the function itself, that denies us full satisfaction and urges us in other directions."

Throughout his writing, Freud also makes clear that the essence of romantic love, the great value the lover places on the loved one, is in fact an illusory overvaluation – a narcissistic extension of one's own ego. The old aphorism that love is blind in Freud's work is elevated to the status of a scientific truth.

James' Monomania View of Love

William James was for many years the dominant figure in American psychology and philosophy and his contributions are indeed mighty in this arena. In his giant compendium on psychology, his sole discussion of romantic love takes place under the rubric 'monomania:'

> The passion of love may be called a monomania to which all of us are subject, however otherwise sane. It can coexist with contempt and even hatred for the 'object' which inspires it, and whilst it lasts the whole life of the man is altered by its presence ([26], vol. 2, p. 541).

James then gives a lengthy, detailed autobiographical case study of a man who became the victim of a blinding passion for a woman from which he managed to extricate himself only after extreme self-induced humiliation. James argues that there is an 'anti-sexual' instinct which repulses us at the thought of bodily contact with another person, even to shaking hands ([26], vol. 2, pp. 437–438).

In James and Freud, two of the dominant psychologists of modern times, we find a cynicism about romantic love that has vastly influenced the field of psychotherapy and sex therapy.

Reich and Adler: NeoFreudian Romantic Love

Wilhelm Reich was a man unafraid of taking an idea to its limits. He built Freud's earliest observations on the role of sexual frustration in the formation of neurosis into an all-encompassing philosophy. Note the personal tone as he tells us:

> My contention is that every individual who has managed to preserve a bit of naturalness knows that there is only one thing wrong with neurotic patients; the lack of full and repeated sexual satisfaction ([36], p. 73).

Faced with this same viewpoint, Freud gave up on the ideal of human happiness and bequeathed the world a philosophy of gloom consistent with his own experience. Not so with Reich, who instead set the goal of achieving human happiness through . . . bigger and better orgasms. When Reich tells us that 'orgiastic potence' is the secret of life, he means it. He is not talking about romantic love for an esteemed other, but pure sexual pleasure, megalomaniacally redefined into a pangalactic 'Force' that makes the Force of Star Wars seem like a cosmic piker. Freud's libido may have been the foundation of human relationships and civilization, but Reich's orgone is the ruling principle of everything in the universe, human and animal, spiritual and physical. It can power space ships, cure cancer, and be seen through a microscope.

Adler [1, 2] was not nearly so imaginative as Reich, far more humble in his personal aspirations toward theoretical grandeur, and definitely more interested in real human beings. His philosophic orientation is interpersonal and social, with a clarifying dose of egalitarian politics.

> If partners are really interested in each other, there will never be the difficulty of sexual attraction coming to an end. This stop implies always a lack of interest; it tells us that the individual no longer feels equal, friendly, and cooperative towards the partner, no longer wishes to enrich his life ([2], pp. 433–444).

This quote bears rereading. It is far ahead of many more contemporary treatises which largely and even wholly ignore love and relationship in evaluating 'sexual dysfunction.'

Havelock Ellis: Advocate of Romantic Love

Ellis wrote in the same period as Freud and William James, the last decade of the 19th Century and the early part of the 20th. Unlike James and Freud, he defied the sexually oppressive attitudes of his generation by finding within himself a love for the human body and for the human beings who inhabit those bodies. In *Studies in the Psychology of Sex* [17] in a chapter entitled 'The Valuation of Sexual Love' Ellis, in partial agreement with Freud, finds that 'the spiritual structure' of life – "our social feelings, our morality, our religion, our poetry and art" – are to some degree built on sex. But for Ellis this enhances the value of sex and sexual love, and culture as well, by integrating humankind's physical existence with its spiritual life.

In *Sex and Marriage* [19] Ellis defines spirituality in a manner consistent with my own usage:

By the term "spiritual" we are not to understand mysterious or supernatural qualities. It is simply a convenient name, in distinction from animal, to cover all those higher mental and emotional processes which in human evolution are gaining greater power. These include not only all that makes love a graceful and beautiful erotic art, but the whole element of pleasure in so far as pleasure is more than mere animal gratification ([19], pp. 35–36).

More explicitly, according to Ellis, "This means that sex gradually [in evolution] becomes intertwined with all the highest and subtlest human emotions and activities, with the refinements of social intercourse, with high adventure in every sphere, with art, with religion."

In *Studies in the Psychology of Sex*, Ellis makes clear that the combination of sex with love can become the model for happiness. Freud made the same conclusion but abandoned all hope for sexual fulfillment – and hence happiness – in the real world. Ellis believed erotic love was both an earthy and earthly reality:

For most people, however, and those not the least sane or the least wise, the memory of the exaltation of love, even when the period of exaltation is over, still remains as, at the least, the memory of one of the most real and essential facts of life ([17], vol. 2, Part III, pp. 137–138).

Fromm's Art of Loving

For Fromm [25] the "deepest need of man" is "to overcome his separateness, to leave the prison of loneliness." Romantic love, which he calls 'erotic love,' attempts to breach the inevitable gulf between individual human beings. It must not be used in a compulsive fashion to deny separateness, but must instead recognize the paradox of the need for both closeness and individuality.

Fromm believes "In essence, all human beings are identical. We are all part of One; we are One." Love for another person then draws upon and inspires love for all human beings. Fromm voices his idealism in conscious opposition to Freud's cynicism.

Fromm parallels love for God with love for people – a theme that appears frequently in authors who value romantic love: "as the logical consequence of theology is mysticism, so the ultimate consequence of psychology is love."

Applying his concepts directly to sexual dissatisfaction and dysfunction, Fromm finds that sexual problems reflect psychological and ultimately spiritual conflicts. Specifically, the emphasis upon sexual knowledge and technique is futile:

Fear of or hatred for the other sex are at the bottom of those difficulties which prevent a person from giving himself completely, from acting spontaneously, from trusting the sexual partner in the immediacy and directness of physical closeness. If a sexually inhibited person can emerge from fear or hate, and hence become capable of loving, his or her sexual problems are solved. If not, no amount of knowledge about sexual techniques will help ([25], p. 89).

The Peak Experience of Maslow . . . and William James

For Maslow [29] peak experiences are "the most wonderful experience or experiences of your life" in which time stands still as we fully immerse ourselves for the moment in total appreciation and enjoyment of an aspect of existence. What interests us here is his inclusion of romantic love as one of the many forms of peak experience, which also includes mysticism, art, and the love of nature. The romantic lover is not deluded into blindness, but instead "is able to perceive realities in the loved one to which others are blind, i.e., he can be more acutely and penetratingly perceptive."

It is fascinating to discover how closely Maslow's peak experience parallels William James' much earlier work on *The Varieties of Religious Experience* [27]. The religious experience for James, including that peak experience called mysticism, is a very personal contact between the individual and God in which the individual's most personal desires find fulfillment. The isolated individual, James tells us, suffers from a profound 'uneasiness' which reflects in its simplest terms "a sense that there is something wrong about us as we naturally stand." The solution is "a sense that we are saved from wrongness by making proper connection with the higher powers."

For James "there is actually and literally more life in our total soul than we are at any time aware of." This unconscious soul or self yearns for perfection. We especially need help in connecting our better nature to a higher spiritual universe, and by loving God we bring out the best in ourselves.

In answer to skeptics, James announces "God is real since he produces real effects." The effects are "life, more life, a larger, richer, more satisfying life." All this can be said of romantic love as well.

Frankl's Existentialism

If we can forgive some dreadful macho slippage, including an argument for a double standard in monogamy, Frankl [20] is inspiring in his willingness to bridge the psychological and the spiritual. He tells us that "there are two ways to validate the uniqueness and singularity of the self." One is through the hard work of developing and implementing one's values. The other is through the 'grace' and 'salvation' of loving and being loved.

In contrast to the cynical views of Freud and James, and consistent with the more optimistic attitudes of Adler, Maslow, Fromm, Ellis, and Norton (ahead) love does not make us blind, but helps us see the "spiritual core of the other person, the reality of the other's essential nature and his potential worth."

Consistent with the view I have stated in my *Psychology of Freedom* – that the body is the vehicle, not the source, of love – Frankl declares:

> Even in love between the sexes the body, the sexual element, is not primary; it is not an end in itself, but a means of expression ([20]. p. 157).

The sexual element merely heightens and helps to express the love itself which reaches from one 'spiritual being' to another. Physical love is essentially spiritual love:

> But for the real lover the physical, sexual relationship remains a mode of expression for the spiritual relationship . . . ([20], p. 158).

In keeping with Adler and Fromm, he locates the source of sexual problems within the quality of the relationship:

> The sexual neurotic no longer fixes his mind upon his partner (as does the lover), but upon the sexual act as such. Consequently the act fails, must fail, because it does not take place 'simply,' is not performed naturally, but is willed ([20], p. 180).

The goal in psychotherapy is to break the 'hapless vicious circle' by refocusing the individual upon the spiritual aspects of the sexual relationship – the sharing of love.

The Individualism of Ayn Rand, Branden, and Norton

No one has written more powerfully than Ayn Rand about the relationship between romantic love and individualism [35, 45], but her concept of love seems more akin to esteem – personal approval of the ethical standards and achievements of another person. In her novels, a character is lovable to the degree that he or she conforms to the author's ethical philosophy in all its details. This competitive, soul-less attitude leads to a vision of lovemaking that is rapacious rather than tender. It may also account for why there are no loved children or pets in her massive novels.

Nathaniel Branden [4, 5] has applied Rand's principles to psychology and in recent years has tempered her hard-nosed individualism with a more general sympathy and understanding for human beings.

In *Personal Destinies* philosopher David Norton [32], a more humanistic individualist, reaffirms the social value of romantic love. Romantic love provides the youthful lover the unprecedented resources to break through his or her own egoism with utter concentration upon another human being. As romantic love matures, love for the other person in all his or her aspects leads to a broadening interest in everything around and pertaining to the other. This love desires "the prosperity and fulfillment of the beloved as a unique and precious enterprise that passion reveals her to be," including her relationships with others and with society.

According to Norton, love is not blind, but deeply seeing. Love unfolds for us "the universality of preciousness in persons as unique individuals." Love "abolishes the anonymous and impersonal common world in order that the world of another person may unfold."

In their introduction to *Philosophies of Love*, Norton and Kille place love in its broadest psychological context:

> Every human act is revelatory, but all are not equally so. Some express the fluid periphery of personality, while others divulge its core. Of those which run to the core it is acts of love which are most telling ([33], p. 5).

Norton and Kille go on to tell us that 'acts of love' are never purely sexual but reflect the individual's lifestory or biography. What a marvelous lesson for therapists!

FEMINISTS ON SEX AND LOVE

In *The Feminine Mystique* [21] Betty Friedan describes the feminine image or ideal as 'passive, weak, grasping dependence.' The feminine woman is expected to find 'security' in a man or boy, "never expecting to live her own life." This leads to a vicious circle that produces so-called sexual dysfunction:

> Sex without self, enshrined by the feminine mystique, casts an ever-darkening shadow over a man's image of a woman and a woman's image of herself. It becomes harder for

both son and daughter to escape, to find themselves in the world, to love one another in human intercourse ([21], p. 281).

Friedan draws heavily on humanistic psychology, including Abraham Maslow, to show that the woman who is personally strong is the most likely to enjoy sex. Here the link between 'sexual dysfunction' and the history of male-female relationships becomes well-articulated. The feminine mystique has led to the destruction of the female identity and the mutual inability of men and women to love each other, sexually or otherwise. Impotence and frigidity are not 'diseases' but responses to the social dysfunctions inherent in modern sexual relationships.

Kate Millet casts an even more harsh judgment upon male values and male sexual conduct. In *Sexual Politics* [31] she provides us a scathing intellectual indictment of patriarchal society, chauvinistic romantic love, and various literary representatives of male ethics, including novelists D. H. Lawrence [28] and Henry Miller [30].

Millet is especially critical of romantic love, which she sees as a male-inspired trap for women. She is surely right that men have been permitted to enjoy sex under every imaginable circumstance, while women have been severely limited to situations of love, and even more so, to marriage. Miller and Lawrence are prototypes of male chauvinism elevated to hero status. But the exploitation of women reflects a general problem within all our social institutions. The challenge is to nurture the essential good in romantic love, while eliminating the elements of power and control as they victimize men and women alike. Romantic love can provide one of the most daring opportunities for men and women to overcome their differences and conflicts through love for one another.

THE QUALITIES OF LOVE

What is love? Whosoever answers that question must first know the answer to 'What is life?' But having established that most good sexual experiences are loving experiences, we must enable ourselves and our clients to identify love. Are we feeling love for our sexual partner? Do we feel love coming from him or her? Have we ever loved or been loved by our partner? Have we experienced love at *any time* in our lives? These questions must be part of any thoughtful examination of sexual issues. We may not need a formal definition of love, but we do need ideas about the attributes of love.

Others have looked in depth at the differences among the various kinds of love (for example, Norton and Kille [33]); but we have thus far found that love can often be discussed as a single entity with only varying degrees of difference between friendship, familial love, love for humankind, erotic or romantic love, and love for God. All forms of love are cut from the same spiritual cloth. If we can say anything about romantic love in particular, it is more intense, more linked to our bodies and hence our passions, more potentially fulfilling.

The Spirituality of Love

Almost everyone who fervently advocates romantic love seems to speak in spiritual terms. Sometimes the spirituality is very secular, but it carries an aura of reverence, a sense that romantic love enables us to see, feel, and experience in ways that transcend ordinary living [7, 19, 20, 32], often connecting us to universal values.

The 'soulness' of all love relationships, romantic or otherwise, has been described by theologian Buber [14] as the 'I-Thou' relationship. Whatever imagery we use to symbolize the most essential aspect or center of the person – soul, being, self, identity, core, heart, depth – most proponents of romantic love believe that the experience reaches from the center of one person to the center of another, seeking divine or fundamental truths in the process.

In its more watered-down versions, love for the person or being becomes love for the person's 'potential.' This emphasis upon potential often misses the mark and gives away the truth that the feeling of love is weak. When we love someone for his or her potential, we do not yet love the person fully, unequivocally, or unconditionally.

Being loved for our potential leaves us hanging, so to speak, with a question mark about our future lovability as we succeed or fail in fulfilling our so-called potential. On the other hand, being truly loved to our core grants us the grace of security; we know we are loved as we are, and for our essence, and not as we might become or as we might perform in the future.

The Omniscience of Love

The lover sees the loved one's essential goodness. When the lover says, 'You are wonderful' or even 'You are the most wonderful woman in the world,' or 'You are perfect,' he is not exaggerating as much as he is proclaiming his vision of the beauty and grandeur within the person. When others cannot see these qualities, it makes no difference; it is their loss. The lover takes joy in knowing the true worth of his loved one. From a practical viewpoint in therapy, when an individual fails to see the beauty and grandeur of another person he or she is probably not in love.

The Unconditionality of Love

Because love reaches toward and touches the essence of the person, love is felt as unconditional and even everlasting [11, 14]. Lovers are like ducklings imprinted forever on their mothers. Once the love is felt it can always be renewed. It may be denied due to hurt and pain; it can be repressed out of a failure to mourn a loss. But it remains within us, ready to be re-evoked whenever we dare. Thus Frankl tells us:

> The existence of the beloved may be annihilated by death, but his essence cannot be touched by death. His unique being is, like all true essences, something timeless and thus imperishable. The 'idea' of a person – which is what the lover sees – belongs to a realm beyond time ([14], p. 154).

Whether or not we agree with 'essences' as a philosophical truth, there is no doubt that the faith in a human essence is one of the hallmarks of romantic love.

Our esteem for a person may rise and fall like a barometer of how we judge the individual's ethical status or achievements. But if this were love, then we could not love infants, pets, plants, or nature itself. Nor could we love grown people when they have behaved in a very disappointing manner. Love pierces through externals, and especially through worldly achievements, touching something that is felt to be real, enduring, and related to the common life force within all of us.

Love Is Joyful

The experience of love is always positive. On this there is remarkable agreement among thoughtful writers. As mystics find joy in their awareness of God, lovers find joy in their awareness of each other. I have described the emotion of love as "joy in the awareness of the existence of another person" [11]. More simply, love is joyful awareness.

Often people who come for therapy have forgotten that love is a joyful experience. When reminded they will usually be able to recall a time in the past when they felt 'in love' and were 'high as a kite' or 'walking on air' or simply 'the most happy I've been in my whole life.' Often that initial love relationship was lost, followed by the setting in of cynicism and skepticism about love. Over the years the joy of love became confused with the pain of loss, betrayal, hurt feelings, jealousy, and guilt feelings. The pain and unhappiness that we usually associate with love are actually the product of the fears and psychological trash we bring along with ourselves into our love.

When I ask a client if she loves her partner, I frequently get this response: 'I get worried if he's sick. I try to help out when he's behind in his work. It upsets me to see him so tired all the time. So I do *care* about him.' Caring is not loving. We can 'care' about anyone, even people we hardly know. Caring is akin to sympathy and a cousin to guilt. Often we experience caring as a burden. It compells us out of guilt to help, support, or sacrifice for people who bring us no joy. Love is joyful, and tends to unburden us as we pursue, perhaps for the first time, our own most cherished desires to be happy.

Love and Fear

Since nearly everyone, including dismal cynics like Freud, agrees that romantic love can bring unexcelled happiness, great fears must block its pursuit. Freud himself, as we noted, spoke about the fear of loss. To this we may add the fear of betraying or being betrayed by the loved one, the fear of controlling or being controlled by the loved one, the fear of needing or being needed by the loved one too much, and so on . . . Since we know that lovers end up seeing deeply into each other, we become afraid of being seen in all our splendor and lack thereof, and of being rejected precisely because of who we are.

For infinite reasons, life is scary. Love draws us into life, often with a force that feels out of our control, and in drawing us into life, it exposes us to every fear of life that we can imagine.

In my therapy practice, among my friends, and in myself, I often have been able to identify the early stages of love by the intensity of the fear associated with it. Frequently a client will tell me about 'this wonderful person I just met' while arguing in the next breath that there's 'no chance at all it can work out.' Usually the individual is making up reasons why it cannot work out in order to protect herself or himself from fears about loving.

The Mutuality of Love

The experience of romantic love dissolves real and imagined barriers between people. At the same time it makes the other person incredibly important to us. The result is the breakdown of distinctions between our own self-interest and the interests of the loved one [11]. When we allow ourselves to act out of love, rather than fear and possessiveness, we want our loved ones to fulfill their hopes and dreams. As the love grows, the mutuality of interests becomes more and more an identity of interests.

When people love each other, giving sexual pleasure becomes a pleasure in itself because we identify so closely with the recipient of our attentions. Giving becomes receiving and the lover becomes grateful for the opportunity to please his loved one.

Love and Need

When we say 'I need you,' we may or may not be talking about love. Sometimes we need people for the fulfillment of unethical purposes and twisted desires. Nor is love the satisfaction of a drive. It is not like hunger or sex where indulgence ends in gratification and loss of appetite.

When lovers have finished making love, they often feel more love, rather than less, and repeated lovemaking can encourage an abundance of loving feelings. Love emanates from our creativity, from our beingness, and hence has boundless opportunities to grow. Sexual satisfaction may put the body at rest, but romantic love liberates our spirit with renewed energies.

Love as Merger

In the desire for merger, romantic love especially parallels mystical union with God. As James found in descriptions of mystical joy [27], sex may color and seem to motivate the experience; but there is a more profound desire for union and oneness with the loved one that far surpasses mere sexual pleasure. This spiritual yearning is rooted in our sense that we are not quite whole without other people, and not perfectly whole without love.

In *The Ways and Power of Love* sociologist Pitrim Sorokin [39] has told us that "the ego or I of the loving individuals tends to merge with and to identify itself with the loved *Thee*," and "More concretely, *love is the experience that annuls our individual loneliness*" ([39], pp. 10–11).

In *Love, Power, and Justice*, theologian Paul Tillich [40] compares romantic love and love for God in regard to the desire for unity with another being. According to Tillich, "Without the desire of man to be reunited with his origin, the love towards God becomes a meaningless word." Between mature, individualized human beings, no real merger is possible, Tillich reminds us; but nonetheless love allows us to transcend our isolation: "Love reunites that which is self-centered and individual. . . . It is the fulfillment and triumph of love that it is able to reunite the most radically separated beings, namely individual persons" ([40], p. 26).

We are, each and everyone of us, isolated human beings, trapped in the shell of our bodies, unable on earth to know another as we know ourselves. Yet in love we may grow to know another better than we know ourselves – in the clarifying light of passionate love. In the instant of love, we know another, and are no longer alone. In an enduring love, the passionate moments grow into a lasting sense that in part at least we are not entirely alone on earth and may share in the joy of knowing and being known. That something as mundane, animalistic, earthy, and instinctual as sex can enhance the experience of loving connection with another human being is one of the puzzles, paradoxes, and miracles of life.

Liberty and Love

'Love brings enslavement.' That's the mistaken formula that governs the lives of many people. Out of fear of being oppressed, controlled, and compulsively attached, many individuals avoid love like the plague. Obviously many of these fears go back to enslaving relationships with parents who manipulated 'love' in order to control their children. Other fears are existential in origin: to love attaches us to life, and all of us have fears of life.

The irony is that we must first find a measure of autonomy and personal freedom in our own lives before we can have the courage to love, and having found the courage, we discover ourselves more attached than ever before. In *The Psychology of Freedom* I put the paradox of liberty and love into these words:

> Just when we begin to feel that we have discovered personal freedom and personal sovereignty, we begin to love. Love then attaches us to people, to places, to objects, to animals, to work, and to ideals. Freedom becomes the right and the capacity to express what we discover within ourselves – our love for others and for life ([11], pp. 218–219).

> Liberty is the context within which each individual can best develop himself or herself. Love is the liberated individual's affirmation of life. Liberty and love are the twin principles of the good and happy life ([11] p. 239]).

TWO MALE 'CASES'

I began this essay with a description of the alternative approaches that a Catholic married woman with 'frigidity' might encounter in America today. The other most commonly cited 'complaint' according to the sex therapy literature is 'impotence' in men, and so it is fitting as we draw near to the conclusion to share two experiences in this area.

A man in his mid-forties came to see me after a year in psychoanalytic therapy in which he had learned a great deal about his negative relationship with his mother, but had accomplished little in overcoming his 'impotence' with his fiancé. His previous psychoanalytic therapy in no way seemed wasted to him or to me, but something was missing.

I asked him early in the session if he felt that he loved his fiancé, and he responded with a description of her so warm and loving that I could feel myself falling in love with her vicariously. I then asked him what connection he made between sex and love. Did he understand that 'having sex' could become the most wonderful expression of the love he felt for his fiancé? No, he was a career army officer who had learned his sexual values from his buddies, and the main value he had learned was something he called 'performance.' He wanted to 'perform' well with his fiancé, for her sake and for his own image.

I suggested to him that he forget about performing, and even about sex itself, and instead concentrate on feeling his feelings of love whenever he lay down next to his loved one. I told him in effect 'If you allow yourself to feel love for her while you are next to her in bed, there is no way you will not become sexually turned on; and if you allow yourself to focus on your love for her while you making love, there is no way you will not stay excited and have a good time.' A little simplified perhaps, but appropriate to his problem, and in two or three sessions he overcame his so-called impotence.

The unusually rapid success was in part due to the achievements of the client's earlier psychoanalytic therapy, and also to his unambivalent love for his fiancé. The next 'case' was more typically complicated.

Joe T. was a director of research at a nationally known institution, and came to me with the mistaken notion that I was a behaviorally oriented sex therapist. He was a scientist by nature and by profession, and conceptualized relationships in terms that would warm B. F. Skinner's heart, if such a heart can be warmed. I explained to him that I had a very different approach based on an alternative philosophy of life. Piqued by my honesty, perhaps, he decided to give therapy a try.

Joe turned out to be a caricature of 'macho male.' He believed that competition, one-ups-manship, and manipulation were at the heart of all human relationship and all human achievement. For years he had experienced 'great sex' with women he met in bars. But when we investigated what he meant by 'great sex,' both his physical and spiritual pleasure turned out to be dismally limited. His major emotion was a fleeting sense of triumph after capturing and mastering an attractive women for a night of sex. The images and sensations reminded me very much of big game hunting, when the hunt is devoid of any love for animals or for the outdoors. Communication and tenderness were absent.

Joe's therapy with me dipped in and out of the whole arena of human life, from humanistic psychology, through feminism, and of course his own particular experience of growing up and living life. By the end of the therapy he had developed his first loving relationship with a woman. In retrospect, he now saw his impotence as an insistence within himself that he did not want to continue in loveless, hostile, and competitive relationships with women in which the only outcome was victory or defeat. His penis was no longer a weapon of triumph over women but an extension of himself for sharing feelings with another human being.

THE MOST COMMON SEXUAL DYSFUNCTION

Sex therapy authorities usually declare that frigidity in women and impotence in men are the two most common complaints or dysfunctions. This is because the truly most frequent problem is such an accepted part of life that professionals tend to overlook it. The most common negative feeling is that 'sex is not all it is cracked up to be' or that it is not as fulfilling as originally hoped. Usually no open complaint is voiced by the sufferer, and instead lifeless sex is accepted as a part of a 'realistic' approach to living.

Sex in isolation from the larger human experience does not have a lot to recommend it for most people. It usually requires alcohol or drugs, violence, temporary infatuation, the allure of danger, or some other artificial device or situation to make it work. But sex as an expression of all that is most satisfying, exciting, and passionate in human relationships can be one of life's most wonderful experiences.

Sex is much more than a bodily drive seeking reduction. Sex becomes embedded in our feelings about ourselves, other people, and life itself. A sexual dysfunction is a life dysfunction and a successful sexual life is inseparable from a good life. At its best, sexual pleasure enhances the joy we feel when sharing life with another human being. This larger understanding of sex enables us to help ourselves and to help others toward a fuller appreciation of life itself.

Center for the Study of Psychiatry,
Bethesda, Maryland,
U.S.A.

BIBLIOGRAPHY

1. Adler, A.: 1969, *The Science of Living*, Anchor, New York.
2. Ansbacher, A. and Ansbacher, R. (eds.): 1956, *The Individual Psychology of Alfred Adler*, Basic Books, New York.
3. Auden, W.: 1948, 'Editor's Introduction', in W. Auden (ed.), *The Portable Greek Reader*, Penguin, New York.
4. Branden, N.: 1963, *The Psychology of Self-esteem*, Nash, Los Angeles.
5. Branden, N.: 1980, *The Psychology of Romantic Love*, J. P. Tarcher, Los Angeles.
6. Breggin, P.: 1971, *The Crazy from the Sane* (a novel), Lyle Stuart, New York.
7. Breggin, P.: 1971, 'Psychotherapy as Applied Ethics', *Psychiatry* **98**, 589–591.
8. Breggin, P.: 1974, 'Therapy as Applied Utopian Politics', *Mental Health and Society* **1**, 129–146.
9. Breggin, P.: 1975, 'Psychiatry and Psychotherapy as Political Processes', *American Journal of Psychotherapy* **29**, 369–382.
10. Breggin, P.: 1979, *Electroshock: Its Brain-Disabling Effects*, Springer, New York.
11. Breggin, P.: 1980, *The Psychology of Freedom: Liberty and Love as a Way of Life*,

Prometheus, Buffalo, New York.
12. Breggin, P.: 1981, 'Madness is a Failure of Free Will; Therapy too Often Encourages It', *Psychiatric Quarterly* **53**, 61–68.
13. Breggin, P.: 1983, 'Iatrogenic Helplessness in Authoritarian Psychiatry', in R. Morgan (ed.), *The Iatrogenics Handbook*, IPI Publishing, Toronto, Ontario, Canada, pp. 39–51.
14. Breggin, P.: 1983, *Psychiatric Drugs: Hazards to the Brain*, Springer, New York.
15. Buber M.: 1958, *I and Thou*, Scribner's, New York.
16. Bulfinch, T.: 1938, *Bulfinch's Mythology*, Modern Library, New York.
17. Ellis, H.: 1936, *Studies in the Psychology of Sex*, vols. 1 and 2, Random House, New York.
18. Ellis, H.: 1939, *My Life*, Houghton Mifflin, Boston.
19. Ellis, H.: 1957, *Sex and Marriage*, Pyramid, New York.
20. Frankl, V.: 1957, *The Doctor and The Soul*, Knopf, New York.
21. Freud, S.: 1930, *Civilization and Its Discontents*, Hogarth, London.
22. Freud, S.: 1961, *Beyond the Pleasure Principle*, W. W. Norton, New York.
23. Friedan, B.: 1965, *The Feminine Mystique*, Dell, New York.
24. Fromm, E.: 1959, *Sigmund Freud's Mission*, Harper, New York.
25. Fromm, E.: 1962, *The Art of Loving*, Harper Colophon, New York.
26. James, W.: 1950, *The Principles of Psychology*, vols. 1 and 2, Dover, New York.
27. James, W.: 1978, *The Varieties of Religious Experience*, Image Books, Garden City, New York.
28. Lawrence, D. H.: 1969, *Lady Chatterley's Lover*, Grove Press, New York.
29. Maslow, A.: 1962, *Toward a Psychology of Being*, Van Nostrand , New York.
30. Miller, H.: 1961, *Tropic of Cancer*, Grove, New York.
31. Millet, K.: 1970, *Sexual Politics*, Doubleday, Garden City, New York.
32. Norton, D.: 1976, *Personal Destinies: A Philosophy of Ethical Individualism*, Princeton University Press, Princeton, New Jersey.
33. Norton, D. and Kille, M. (eds.): 1983, *Philosophies of Love*, Rowman and Allanheld, Totowa, New Jersey.
34. Rand, A.: 1943, *The Fountainhead*, Bobbs-Merrill, New York.
35. Rand, A.: 1957, *Atlas Shrugged*, Random House, New York.
36. Reich, W.: 1942, *The Discovery of Orgone*, Farrar, Straus & Giroux, New York.
37. Rieff, P.: 1961, *Freud: The Mind of the Moralist*, Doubleday, Garden City, New York.
38. Rohr, R.: 1982, 'Toward Sexual Wholeness', *Sojourners* **9**, 14–19.
39. Sorokin, P.: 1967, *The Ways and Power of Love,* Henry Regnery, Chicago.
40. Tillich, P.: 1960, *Love, Power, and Justice*, Oxford University Press, New York.

NOTES ON CONTRIBUTORS

Fritz K. Beller, M.D., is Professor and Chairman, Der Universitäts Frauenklinik, Münster, West Germany.

Peter Roger Breggin, M.D., is a psychiatrist in private practice, and Director, Center for the Study of Psychiatry, Bethesda, Maryland.

Vern L. Bullough, Ph.D., is Dean, Faculty of Natural and Social Sciences, State University College, Buffalo, New York.

Eli Coleman, Ph.D., is Assistant Professor, and Associate Director, Program in Human Sexuality, Department of Family Practice and Community Health, Medical School, University of Minnesota, Twin Cities, Minnesota.

Sandra Harding, Ph.D., is Associate Professor, and Director of Women's Studies, Department of Philosophy, University of Delaware, Newark, Delaware.

Virginia E. Johnson-Masters, D.Sc. (Hon.), is Director, Masters and Johnson Institute, St. Louis, Missouri.

Robert C. Kolodny, M.D., is Medical Director, Behavioral Medicine Institute, New Canaan, Connecticut.

Stephen B. Levine, M.D., is Director, Sexual Dysfunction Clinic, School of Medicine, Case Western Reserve University, Cleveland, Ohio.

Leslie M. Lothstein, Ph.D., is Director of Psychology, Institute of Living, Hartford, Connecticut.

Joseph Margolis, Ph.D., is Professor, Department of Philosophy, Temple University, Philadelphia, Pennsylvania.

William H. Masters, M.D., is Chairman, Masters and Johnson Institute, St. Louis, Missouri.

Jerome Neu, Ph.D., is Professor and Chairman, Board of Studies in Philosophy, University of California, Santa Cruz, California.

Earl E. Shelp, Ph.D., is Research Fellow, Institute of Religion, and Assistant Professor, Department of Community Medicine, Baylor College of Medicine, Houston, Texas.

Alan Soble, Ph.D., is Assistant Professor, Department of Philosophy, St. John's University, Collegeville, Minnesota.

Robert C. Solomon, Ph.D., is Professor, Department of Philosophy, University of Texas, Austin, Texas.

Frederick Suppe, Ph.D., is Chairperson, Committee on the History and Philosophy of Science, University of New Orleans, New Orleans, Louisiana.

INDEX

abortion 19, 97–98, 104
abstinence 5
Acquired Immune Deficiency Syndrome 13, 112, 145
Adler, Alfred 252, 255
amniocentesis 104
Ardrey, Robert 193
Aristophanes 213, 222
Aristotle xxvii, 74, 75
astrology 75–76
Auden, W. H. 247

Barnes, Barry 186
Barth, John 214
Bates, J. 64
Bayer, Ronald 12, 28
Beach, F. W. 225
Bell and Weinberg 230, 234
Beller, Fritz K. xxviii, 87–108
Bem, S. 229
Bentham, Jeremy 127
Berger, Brigitte 193
Berger, Peter 193
Bieber, Irving xxix, 26–28, 47, 117–119
biological determinism 187
biological sex 216
bisexuality 225–241
Blach, Iwan 79
Bloor, David 186
Blumer, D. 58
Blumstein, P. W. 227, 230, 236
Bode, J. 233
Boorse, Christopher xxix, 122, 124, 126–128, 146–149
Bozett, F. 236
Branden, Nathaniel 247, 256
Breggin, Peter xxx, 243–267
Brown, John 78
Brown, Norman O. 165
Buher, Martin 258
Buck, T. 65
Bullough, Vern xxvii, xxxiii, 19, 73–85

Callahan, Daniel 114
Chaddock, Charles G. 23
chimera 91
Clark, Thomas 116
Coleman, Eli xxx, 225–242
Comfort, Alex xii
consensual sexual act xx
contraception xvi, 19, 92–97
Corner, George W. 81
cultural contructionism 187
cunnilingus xiv, 159

Dana, Charles L. 23
Davis, Katherine B. 81
de Beauvoir, Simone 185
DeCecco, J. P. 228–229
D'Emilio, J. 28
Dennett, Dan 210
Derrida, Jacques 207
de Sade, Marquis 214
Dickinson, Robert Latou 81–83
disease 141–150
Don Juan 211, 248
DSM-III 6, 7, 9, 17, 29, 56
Duffy, John xxvi, 17, 19, 25
dysfunction, sexual 4, 7–8

Eber, M. 61–62, 66
Edwards, Susan 129–130
ego-dystonic disorders 31, 237
Ehrhardt, A. A. 58, 59
Ellis, Havelock xxviii, 10, 23, 26, 79–80, 82, 154, 252–253

Elshtain, Jean Bethke 193
Engelhardt, H. T., Jr. xxix, xxxi, 19, 121–123

Fabrega, Horatio xxix, 141–142, 145–146
family dynamics 64–65
Feinberg, Joel 120–121
fellatio xiv, 159
feminism 185, 191–196, 257
Fere, Charles 79
fetishism 7, 23, 161, 170–174
Fithean, Marilyn 83
Ford, C. S. 225
Foucault, Michel 194, 211
Frankel, Victor 255
freedom 132–134, 263
Freud, Sigmund xxix, 4, 11, 23, 80, 127, 132, 153–168, 170–173, 175–182, 210, 212–213, 219–224, 226–228, 250–252, 260
Friedan, Betty 193, 244, 257
Friedman, Richard 116
frigidity 243–246
Fromm, Erich 250, 253–255
function 143–144, 147–150, 212–215

Gagnon, J. H. 226
Galen 73–74
gay liberation 28
gender identity 29, 33, 40–42, 47, 144, 216
gender-role stereotypes 31
geriatrics 3, 4
Gilder, George 193
Goldman, Alan xi, xiii, xx
Gorer, Geoffrey xv
Gould, Robert xxix, 116
Graham, D. L. R. 234
Gray, Robert xx, 125
Green, Richard xxix, 117–118, 128

Haller, J. S. 19
Haller, R. M. 19
Hamilton, Allen McLane 23
Hamilton, G. V. 81
Harding, Sandra xxix, 185–203
Hardwick, Elizabeth xii
Hartman, William 83
Hatterer, L. 13
health, sexual 3, 8–11, 111–134, 140–141
Hellegers, Andre 92
heterosexuality xxii, 205–224
Hirshfeld, Magnus 4, 23, 80, 82
homoeroticism 48
homophobia 13, 28
homosexuality xvii, xviii, xix, 11, 27, 46–49, 89, 123–125, 155–157, 165–170, 230
Hooker, Evelyn 11, 27
Humphreys, Laud 230

impotency 28, 73, 81
infantile sexuality 162
infertility 98–102
inhibited sexual desire 6
intersubjectivity 222–223
in vitro fertilization 105–106
Johnson, Virginia ix, xii, xxviii, xxix, 3, 5, 9, 12–13, 83, 117–118, 125
Illich, Ivan 129, 193

Jagger, Alison xv, xvi
James, William 252, 254–255, 262
Jorgensen, Christine 29

Kant, Immanuel 213, 218
Kaplan, Helen Singer 8, 9, 143
Katz, J. N. 22
Kiernan, James G. 23
Kille, M. 256–258
King, C. Carly 147
Kinsey, Alfred xiv, xv, xvii, xxviii, 10, 26, 28, 33, 82–83, 225, 227–231, 240
Klein, Fred 226, 231–232, 235, 239
Kohlberg, Lawrence 127
Kohut, H. 67
Kolodny, Robert C. xxvi, 3–16
Kovel, Joel 127
Kraft-Ebbing, Richard Von xxviii, 79, 80, 154

Lasch, Christopher 193
LaTorre, R. W. 234
Latour, Bruno 186
Lawrence, D. H. 206, 214, 257
Leiss, William 194
lesbianism xviii, 47–48
Lester, D. 25
Levin, Max xxix, 115–116, 127
Levin, Michael 193
Levine, Stephen xxvi, xxvii, 39–54
Limentani, A. 63
Litin, E. 64
Lo Piccolo, J. 8
Lorenz, Konrad 193
Lothstein, Leslie xxvii, 55–72
love 23, 114, 218, 222, 247–265
Lumiere, Richard 112

Mahler, M. 62
Marcuse, Herbert 165
Margolis, Joseph xxix, 122–123, 139–152
Marmor, Judd xxix, 27, 116
Martin, C. W. 225
Maslow, Abraham 254, 257
Masters, William ix, xii, xxviii, xxix, 3, 5, 9, 12–13, 81, 83, 117–118, 125
masturbation xi, 17, 18–34
May, Rollo 221
McCarthy, Joseph 27
McDonald, A. P. 230, 241
Mead, Margaret 225, 236
Meyenberg, B. 66
Meyer-Bahlburg, H. 59
Meyer, J. 62, 65
Mill, John Stuart 131
Miller, Henry 257
Millet, Kate 244, 257
Mill, Albert 4, 79
Money, John 58, 142, 216
Morgan, Robin xxiii
Morris, Desmond 193
Mosher, Clelia 81

nature – the 'natural' 5, 80, 82, 122–127, 140, 185–200, 208, 213–214

Neu, Jerome xxix, 153–184
Newman, N. 65
Nietzsche, Friedrich 207, 209
normality – the 'normal' 32, 140
Norton, David 256–258

O'Brien, Mary 195
Ovesey, Lionel 31, 66

paraphilias 28–29, 43
Paul, J. P. 231
perversion 23–25, 28, 51, 62–63, 79, 129–130, 153–176
plastic surgery 31
politics of sexuality 191–196, 228
Pope Paul VI xxiii
pornography 115
Pauly, I. 64
Person, E. 66
Plato 139, 222
Pomeroy, W. 225
pregnancy 73–74
prejudice 22
Prince, V. 30
procreation 22, 149, 218
psychiatry 23, 26, 32, 81, 112–113, 175
psychosexual dysfunction 28, 31

Rado, Sandor 11–12
Rand, Ayn 256
Rand, C. 234
rape xixff
Ravetz, Jerome 186
Rawlings, E. I. 234
Rawls, John 127
Reich, Wilhelm 132, 252
reproduction xiii, 164, 218
reproductive technology 87–107
Rich, Adrienne 195
Rieff, P. 250
Rohie, W. F. 82
Robins, Eli 11–12
Rohmer, Eric 211
Rohr, Richard 244
Rossi, Alice xv, 193
Rousseau, Jean-Jacques 220

Rudolph, J. 229
Ruse, Michael xxix, 124–125

Saghir, Marcel 11
Sanger, Margaret 94
Sartre, Jean-Paul 211, 222–223
Schwartz, P. 227, 230, 236
sex reassignment surgery 29, 55, 57, 69–70
sex research 26–32, 73–83
sex-role stereotypes 23, 229
sex therapy 14, 73–83
sexology 22, 26
sexual disorders ix, 49–51, 117–120
sexual ethics 217
sexual feelings 4, 211, 212
sexual identity 33, 40, 228
sexual intention 40, 42–44
sexual norms – medical and biological 26–34
sexual norms – moral 78–79, 111–112
sexual orientation 12, 33, 40, 42, 216
sexual pleasure xxi, 219–222
Shelp, Earl E. xxv–xxxii
Shively, M. 228–229
Sieguesch, V. 66
sin xi
Skinner, B. F. 264
Slater, Philip 206
Soble, Alan xi–xxiii, xxviii, xxix, 111–138
Socarides, Charles xxix, 11, 28, 47, 62–63, 118
Socrates 139
Sohn-Rethel, Alfred 186
Solomon, Robert xxx, 205–224
Sorokim, Pitrim 262
spirituality 258
Spitzer, Robert 116
St. Albertus Magnus 77
Stoller, R. 60–65, 68
Storms, M. D. 229
Suppe, Frederick xxvi, 17–37
Szasz, Thomas 13, 112, 148

Tatum, H. 94
teleology 207–209
Thoinot, L. 79
Thrasymachos 139
Tiger, Lionel 193
Tillich, Paul 262
Tissot, S. A. 10, 77–78
transsexualism 23, 30–31, 41, 55–70
transvestism 23, 41

Van deGelde, Theodore 10
Veatch, Henry xxix
Veatch, Robert 113, 145–146
venereal disease 19, 112
Victorian sexual attitudes 22–25, 32
Volkan, V. 66

Wayson, P. D. 232–233
Weinberg, M. S. 230, 233
Wendenberg, K. 234
William of Saliceto 76
Williams, C. J. 230, 233
Wilson, E. O. 193
Wolf, Charlotte 231
Woolgar, Steve 186

The Philosophy and Medicine Book Series

Editors

H. Tristram Engelhardt, Jr. and Stuart F. Spicker

1. **Evaluation and Explanation in the Biomedical Sciences**
   1975, vi + 240 pp. ISBN 90-277-0553-4
2. **Philosophical Dimensions of the Neuro-Medical Sciences**
   1976, vi + 274 pp. ISBN 90-277-0672-7
3. **Philosophical Medical Ethics: Its Nature and Significance**
   1977, vi + 252 pp. ISBN 90-277-0772-3
4. **Mental Health: Philosophical Perspectives**
   1978, xxii + 302 pp. ISBN 90-277-0828-2
5. **Mental Illness: Law and Public Policy**
   1980, xvii + 254 pp. ISBN 90-277-1057-0
6. **Clinical Judgment. A Critical Appraisal**
   1979, xxvi + 278 pp. ISBN 90-277-0952-1
7. **Organism, Medicine, and Metaphysics**
   Essays in Honor of Hans Jonas on his 75th Birthday, May 10, 1978
   1978, xxvii + 330 pp. ISBN 90-277-0823-1
8. **Justice and Health Care**
   1981, xiv + 238 pp. ISBN 90-277-1207-7
9. **The Law-Medicine Relation: A Philosophical Exploration**
   1981, xxx + 292 pp. ISBN 90-277-1217-4
10. **New Knowledge in the Biomedical Sciences**
    1982, xviii + 244 pp. ISBN 90-277-1319-7
11. **Beneficence and Health Care**
    1982, xvi + 264 pp. ISBN 90-277-1377-4
12. **Responsibility in Health Care**
    1982, xxiii + 285 pp. ISBN 90-277-1417-7
13. **Abortion and the Status of the Fetus**
    1983, xxxii + 349 pp. ISBN 90-277-1493-2
14. **The Clinical Encounter**
    1983, xvi + 309 pp. ISBN 90-277-1593-9
15. **Ethics and Mental Retardation**
    1984, xvi + 254 pp. ISBN 90-277-1630-7
16. **Health, Disease, and Causal Explanations in Medicine**
    1984, xxx + 250 pp. ISBN 90-277-1660-9
17. **Virtue and Medicine: Explorations in the Character of Medicine**
    1985, xx + 363 pp. ISBN 90-277-1808-3
18. **Medical Ethics in Antiquity: Philosophical Perspectives on Abortion and Euthanasia**
    1985, xxvi + 242 pp. ISBN 90-277-1825-3
19. **Ethics and Critical Care Medicine**
    1985, xxii + 236 pp. ISBN 90-277-1820-2
20. **Theology and Bioethics: Exploring the Foundations and Frontiers**
    1985, xxiv + 314 pp. ISBN 90-277-1857-1
21. **The Price of Health**
    1986, xxx + 280 pp. ISBN 90-277-2285-4